RADIAL IMPLICATIONS
OF THE UNIFIED FIELD

RADIAL IMPLICATIONS OF THE UNIFIED FIELD

Classical Solutions for Atoms, Quarks and Other Sub-atomic Particles

JONATHAN O. BROOKS

RADIAL IMPLICATIONS OF THE UNIFIED FIELD
CLASSICAL SOLUTIONS FOR ATOMS, QUARKS AND OTHER SUB-ATOMIC PARTICLES

iUniverse books may be ordered through booksellers or by contacting:

iUniverse
1663 Liberty Drive
Bloomington, IN 47403
www.iuniverse.com
1-800-Authors (1-800-288-4677)

ISBN: 978-1-4917-7535-6 (sc)
ISBN: 978-1-4917-7536-3 (e)

Library of Congress Control Number:2015914102

Print information available on the last page.

iUniverse rev. date: 10/07/2015

Contents

Contents

Preface

As a Chemical Engineer from Purdue at Pfizer I was given the problem of separating two closely similar chemicals. It had been previously determined that two non-aqueous solvents constituted the best system. I derived an *equation* that showed my employers that there was a critical concentration for the process else premature crystallization would destroy the separation.

In the early sixties I was using my slide rule to examine the *equation* that I had derived. The digits 529 came under the hairline of my slide rule. This, I recognized, was indicative of the radius of hydrogen. I found that the radial equation was a solution to the Bessel type equation that I had earlier derived for the separation of two similar chemical substances. Dr. Deakins of the Indiana State University confirmed that the equation that I had derived was indeed a Bessel equation. From it I was able to find aufbau matrices from which I could replicate known radii of all elements of the Periodic Table. The *1s* parameter of these radii formed a periodicity with atomic number that was an inverse of the often presented periodicity of radii. I taught secondary school mathematics, chemistry and physics for thirty nine years. I earned a PhD during summers at Indiana State University. After retirement I did substitute teaching and taught physics for seven years at a junior college.

For over 50 years I communicated my most recent results to the MAA, the AAPT, the ACS and members of the Nuclear Division of AIChE. Their patience with me I sincerely appreciate. Every part of this work which is a success is attendant with one or more approaches that were not so successful. Many lost interest in my talks I'm sure.

I hope that this book will reach those who are in a position to profit from these writings. There no intent to deprecate the work of many able persons. There are a half dozen *ideas not supported*. I first will list over a *dozen of new items* that I believe are true accomplishments that are not now found in the current literature: (1) A solution to the radial portion of Schrödinger's equation; (2) A method for calculation of the Poisson applicable to both mass and electricity; (3) A mechanism for verifying the published radii of the atoms by a variable I call the attribute which is unique to each atom; (4) A showing of the variation with temperature of the foregoing attribute; (5) Proof of Einstein's conjecture that gravity would be found to be an electromagnetic phenomena: (6) Calculation of the spectra of hydrogen and helium with concomitant attachment of the radial status after each photon enters; (7) Suggestion that attachments of electron from supernovae can alter the apparent value of Newton's Universal Gravitation Constant G which would create some "dark matter": (8) Finding a viable orbital system for quarks and electrons; (9) Showing that protons and neutrons have their mass as a result of an orbital system; (10) Revision of some usual mechanism of the weak force; (11) Working in the coulomb gauge throughout smallest to largest domains; (12) Last, but not least is the enunciation of the *unified field* which enabled the forgoing to happen not in phase space but *in ordinary space*. Together the above theory can be called a TONE with NE signifying "Nearly Everything".

There are nearly an equal number of phenomena, now much used, but not supported by my findings. Most, but not all, of the players who enunciated or espoused these ideas are not living. I hope that those who are living and working with these ideas will rally and take quantum attributes forward. The mathematics as used are primarily algebraic. Complementarity, the Correspondence Principle and Uncertainty are not herein supported. Likewise, de Broglie's hypothesis, and Noether's theorem as applied are not tenable not only because the photon has no mass, but also because we are dealing with a Newtonian system; hence, despite Planck's constant, there is no momentum to transfer. The photon does have energy which is always conserved. Angular momentum is always conserved *within* the atom, even in elliptical orbit. The Lagrangian as used by quantum mechanics and Einstein in General Relativity is of *validity in only the cosmic sense*, as of a falling ball, and hence is not the reason that the "Holy Grain" of Physics was enunciated to herald the incompatibility of the two. The real culprit was the ambiguous gauge discrepancy. With now no gauge discrepancy because we can work entirely in real space, not phase space, the grail loses its significance between the small atomic, subatomic and the solar domains. This raises now some question of the significance of the Standard model.

Lest anyone comes to believe that chapter one is but a mathematical artifact I offer two facts to dispel that idea. First the ideas came from the rearrangement of the Rydberg equation. Second the associated magnetic phenomena agree with data that NASA took in their flyby of the planets. At the same time the idea that a real non-integral number may just be the quantum gravity of our solar system and that a time-like variable N may resurrect, albeit weakly, Newton's idea of a universal time because it defines a variable alpha that starts at a zero time and runs to one is controversial. A closer look at Bohr's derivation of the Rydberg constant will show that the kinetic energy portion is Insert Figure 3-10 not present; although it plays a prominent part in the Schrödinger equation.

I used a version of that portion of the Rydberg rearrangement to find supposed solar quantum numbers earlier; but, I omitted the 2 and used mean solar velocities. I found the set of solar quantum numbers for the nine planets as {1.08, 1.45, 1.70,

2.08, .3.88, 5.76, 7.44, 9.33, and 10.69}. Consider N as a candidate for universal time. Observe our time for the time of revolution of a planet about its central mass. Using an earth N of 1.70 and the ratio of our time of revolution to our time of their revolution calculate an N for their planet. Using the formula for N, *N h V=G q,* calculate their *V*, and with our time and their V we can calculate a radius. With a radius and a velocity we can calculate a mass for them by Newton's Law. One catch is that even if their N is different will its use calculate their velocity? I know that it does not give the mass of our Sun; although, the N's of our other planets do. There velocity is keyed to their N. So it appears that we justified in using our units even if their supposed inhabitants had their own. Can we conclude that every planet has a universal N with a mass in our unit of a kilogram? There are an infinity of real numbers between zero and eleven.

Unfortunately I had anticipated an increase in the base radius of hydrogen and helium as energy was added for a number of years before I realized that the radius must decrease as energy is added which sustains the electron in orbit. Later chapters extend the radial equation and orbital action to the sub-atomic domain. This will replace the Lagrangian and Hamiltonian versions of quantum mechanics practice. These are the ideas embellished in phase space which render quantum mechanics inconsistent with general relativity. My idea of applying classical unified fields to particles of the zoo is new.

The older quantum theory had a discontinuity in the mechanical portion of the quantum process. Quantum mechanics arose as a separate discipline because there was a disharmony between the undulatory and corpuscular theories of light.

Jammer[1], in explaining the dilemma at that time, relates that A.H. Comptons's experiment had confirmed Einstein's heuristic viewpoint regarding Planck's quantum hypothesis. Similarly, Fresnel's undulatory theory had explained the Young's experiment. Louis de Broglie, the author of quantum dependence of angular momentum, supported Compton. Schrödinger, initially a pledged advocate of the wave theory, gave a quantum proof of the highly undulatory Doppler Effect.[2] Still diffraction and interference resisted quantum explanation. It is notable that these are *position and velocity effects* of the electron and photon that are *exterior to the atom.* Describing these phenomena required measuring wavelengths. However, the light portion of the quantum process was continuous. It was not appreciated then that the photon nature of light was interrupted upon reaching the atom *adding only its energy* thereto, but not angular momentum.

---

[1] Jammer, Max, Conceptual Development of Quantum Mechanics, McGraw-Hill, New York, © 1966, p.157f.

[2] Schrödinger, E., "Dopplerprinzip und Borsche Frequenzbedingung", *Physikalische Zeitschrift*, 23, 1926 p. 301ff.

We shall see that measurement of change in velocity in orbit is sufficient to characterize only one unit of the numbers of quanta, n, called quantum numbers, at a time. The energy in the entering photon allows us to find a change in radius to represent the energy imparted to test object. I hope that it is now evident that the controversy over the particle-wave nature that spawned the advent of Quantum Mechanics is resolved by saying the photon is a wave and the atom with its electron is a composite particle.

Slater became convinced of the one electron, one quantum idea established by Compton. He first tried to reconcile quanta with waves by coupling a Poynting vector to the magnetic field energy density. Magnetic energy density transforms as a vector and would have represented the velocity imposed by a quantum. He later concluded that the only way to proceed was to find a statistical connection between light quanta and waves. The main connection between spectra and waves is the two relations $v = f\lambda$ and the ubiquitous $E = hf$. Frequency as such a connection became d Broglie's thesis. Since one needs *a velocity or a quantum number* to complete the description of the orbit, it would appear that general relativity does indeed have a velocity and is fully complete and tentatively accepted. Because on closer examination what constitutes the quantum is really a *specific energy*, which in cosmology, is supplied by a central mass holding a smaller mass in orbit. Reflection will show that if quantum has to refer to the value of a variable n then there is no quantum theory. Since quantum refers to the amount of energy let us be happy and say we have an addition to quantum mechanics to present in this book. Why should not this energy be conditioned by the Poisson which will explain "dark matter"? And, why if the central mass becomes clouded enough with negative charge would there not be a "dark energy".

My genealogical roots extend from my father's side to the Anglo Saxons and from his mother, my paternal grandmother, through Admiral Thomas Hardy and three other British Admirals Berkeley, a family originally from Denmark. On my mother's side, her mother traces from England, Italy and her father's ancestry is from two American Revolutionary soldiers one whose family came here from Bavaria in the early 1600's. These are the countries that contributed most to quantum theory.

The successes of Quantum Theory are largely where waves of heat or electricity are propagated through macroscopic matter. Quantum attribute theory is a bottoms up theory from the electron that Quantum Theorists will either advance or possible decline to advance because it controverts quantum mechanics of phase space. Both theories have their respective domains. Any one reading this book will find that there is much yet to be done to advance this theory where quantum mechanics has not gone. One critical area is in the attachments of medicines to their targets. This is an area where the relative unified fields of atoms is critical in defining potentials. Another critical area of need is in the rare earth modification of solar cells.

Dedication

This book is dedicated to my parents, Oswald, and Lillian, Maud, (Yeager) Brooks, my wife, Mafalda (Fenoglio) Brooks, and my daughter Pamela Marie, all deceased.

I want to include my now constant companion, Mary Lou (Buckingham) Santus, who was my classmate from the seventh grade through High School in this dedication.

Chapter One

Unified Field of the Solar System

I write concerned with providing a physical justification that electric, magnetic and gravitational forces are all due electromagnetic phenomena. This is true for gravitational at macroscopic level for celestial bodies consistent with relative permeability and permittivity. In 1953 Albert Einstein wrote to the Cleveland Physics Society on the occasion of a commemoration of the Michaelson–Morley experiment. In that letter he wrote[3] "The fact led me more or less directly to the special theory of relativity was the conviction that the electromotive force acting on a body in motion in a magnetic field was nothing else but an electric field." Einstein is said to have asserted that gravity was electromagnetic. However, in his latter days he was unable to prove it. It is my belief that this chapter will detail such a proof. Although, at microscopic level (atomic level and below) attempts to reconcile or unify electromagnetic and gravitational forces based on quantum and classical mechanics have ostensibly been shown to be viable[3], very little or no effort has been made[4] to show that the various forces, both electric and magnetic, can be reconciled with Newton's laws at macroscopic level (astronomical level). Such unification at sub solar levels generally assumes that gravity is negligible and that is not correct.

Declaration of Maxwellian Gravity:

Preliminary data on charge equivalence to gravity has been previously given in reference.[4, 5] It is shown here that the similar results are achievable for magnetic fields of the planets and sun. The impetus for describing gravity by Coulomb's law comes from rearrangement of a portion of the equation as derived by Niels Bohr to exemplify the Rydberg constant:[5]

---

[3] 3. R. S. Shankland, Am. J. Phys. 32 (1), 16–35 (1964).

[4] A. Michaud, The General Science Journal, January 17 2011: Unifying All Classical Force Equations.

[5] J. O. Brooks, *Quantum Gravity and the Unified Field*, Vantage Press, Inc. NY, © 2000 p. 18 ff.

[6] ibid p 66.

$$\frac{1}{\lambda} = \frac{q^4 m}{8\varepsilon_o h^3 c}\frac{1}{n^2}$$

1-1 a

In equation (1a), λ is the wavelength, q is charge on electron, m is the mass of an orbiting entity, ε_0 is the permittivity of free space, c is the speed of light, h is Planck's constant, n is a unit quantum number, and m is the mass of electron. Introducing, Coulomb's law proportionality constant $K = 1/4\pi\varepsilon_0$, $\hbar = h/2\pi$, M as the central mass, m is the test mass, Q is charge of the proton, q is the charge on the electron, and then r is the distance between the two charge bearing masses. Finally, v is the velocity of electron. Upon rearranging equation (1-1 a) then dividing by Mm, we find, on each side, the units of Newton's universal gravitational constant, G.

$$\frac{n\hbar\sqrt{cv}}{Mm} = \pm\frac{KQ\sqrt{\frac{mvr}{\hbar}}}{Mm}$$

1-1b

$$\frac{n\hbar V}{q} = G \qquad \frac{G}{r^2} = \pm\frac{KQq}{Mmr^2}$$

1-1 c d

In above equation (1-1 C), V is the orbital velocity of planet around the sun. For present we ignore this LHS of above equation which leads to quantum gravity; and, the negative sign of the RHS now divided by the radius squared which in another article[7] will be shown to lead to the unified atomic gravitational and electromagnetic fields. In equation (1 b) as the units on each side are those of G, we can simply equate the RHS to G. The radical is set equal to plus or minus unity as per unit quantum jump; therefore, multiplying by M and dividing by the square of radius we have the units of the local or planetary gravitational constant, g.

$$\frac{GM}{r^2} = +\frac{KQq}{mr^2} = g$$

1-2

Note that in order to obtain equality the coulombic charges Q and q can no longer be that of the charge on an electron or a proton. By fitting the known planetary gravitational fields as the g of equation (1-2) and assuming the positive charge on a protons of the sun we can determine the negative charge on the planets. We find that while the negative charge on a planet *is not constant* the negative charge per unit of planetary mass *is constant*. In short, there is a neat way to arrive at this.

We have found that $G = KQ q$, wherein Q is defined as the product of Avogadro's number (N_{AV}) and the calculable positive charge on a kilogram of proton mass. The negative charge q is our newfound experimental charge on a kilogram of negative mass where with we reproduce the gravitational laws.

$$Q = 6.0221450 \times 10^{26}\frac{atoms}{Kg} 1.602217765 \times 10^{-19}\frac{coulomb}{atom} = 9.64878138 \times 10^7\frac{C}{kg}$$

1-3

Next we obtain q in terms of K, Q, and G:

$$q = \frac{G}{Kq} = \frac{6.673848x10^{-11}}{(8.98755179x10^9)(9.64878138x10^7)} = 7.69595387x10^{28}\,C/kg \qquad \text{1-4 a}$$

It is convenient at this point to define a constant, Θ:

$$\theta = \sqrt{\frac{G}{K}} = \sqrt{Qq} = 8.617225394x10^{-11}\,C/kg \qquad \text{1-4 b}$$

Both Q and q are on the basis of charge per kilogram so that we can write both the local and planetary gravitational field equations representing KQq or $K\Theta$ as G with a mass term and a radius squared. The values obtained in equation (3) and (1-4 a) are essentially the same. Similarly: se evaluated in reference[4, 5] but in present work utilizes the most recent values of the constants involved.

The equation (2) may now be rewritten for local gravity as

$$g = \frac{GM}{r^2} = \frac{KqQM}{r^2} \qquad \text{1-4 c}$$

$$\frac{GM}{r^2} = \frac{8.98755179x10^9)(9.64878138x10^7)(7.69595357x10^{-29})M}{r^2} \qquad \text{1-5 a}$$

This is necessary since the charges are both on a unit kilogram. Alternately, we may write: as:

$$g = \frac{(8.98755179x10^9)(8.617225394x10^{-11})M}{r^2} \quad \text{or} \quad g = \frac{K\theta^2 M}{r^2} \qquad \text{1-5 b}$$

The mass on the right in equations (5 a, b) can be either the central mass of the sun or the central mass of the planet. Since the coulombic values are both per unit mass basis the requisite Newton per kilogram expresses the gravitational field. If the mass of the sun is used the total positive charge of the sun is given by (Qx Mass of Sun) and the test charge will have to be (q) for each planets at a distance, r from the sun. If the mass of a planet is input for M then Q is the test charge and r is the radius of the planet. As can be seen from Table 1 that the calculated gravities both on the planets and sun as evaluated with the Coulomb's law reproduces Newtonian planetary dynamics without any divergences in numerical values. Since, $K\Theta^2 = G$, one may speak that we have a correlation coupling gravity to Maxwell's electromagnetic theory. We now seek to extend the correlation to the magnetic portion of that theory. The Table 1-1 gives the reevaluated and updates previous values for solar charge and gravities for planets in our solar system.

As an example of how the gravity values are calculated in terms of charges Q and q; the author's feel that readers should be made aware that following format was used originally: r stands for radial distance between central and test charges. For gravity product of the suns Mass and Q C/kg is the central charge and q represents the negative test charge per kg of planet. For local gravities, the product of planetary mass and q represents the negative central mass and Q becomes the positive test charge orbiting the planet.

It can be seen from the following Table 1 that calculated values for gravity both on the planets and sun as evaluated with the Coulomb's law reproduces Newtonian planetary dynamics without any divergences in numerical values. In dealing with local gravity we can rewrite equation (5 b) using Coulomb's law constant in the following form (where M can be either the mass of planet or Sun

$$g = \frac{GM}{r^2} = \frac{\theta^2 M}{4\pi\varepsilon_o r^2} \qquad \text{1-6}$$

Using the definition of μ_0 we can write:

$$g = \frac{c^2 \mu_o \theta^2 M}{4\pi r^2} \qquad \text{1-7 a}$$

We can now define magnetic field:

$$B = \frac{c \mu_o \theta^2 M}{4\pi r^2} \ \text{T} \qquad \text{1-7 b}$$

In above equation (1-7 b), B has units of $NA^{-1}m^{-1}$ (or Tesla) and the value of Θ has been already defined by equation (1-4 b). It is evident that equation (1-7 a) will give identical result for local and planetary gravities as was found for each case of Coulomb and Newton laws.

Table 1-1 Planetary Variables

Planet	Mass m (x10^{24}) kg	mxq (C)	Q (x10^{-29}) (C/kg)	Distance to Sun, r (x10^9) (m)	Qx10^7 C/kg	MxQ (x10^{38}) (C)	g at top (m/s^2)
Mercury	0.330	2.54x10^{-5}	7.696	57.9	9.648	1.919	3.7030
	4.868	3.746x10^{-4}	7.69	108.2	9.648	1.919	8.871
Earth	5.973	4.597x10^{-4}	7.696	149.6	9.648	1.919	9.8004
Mars	0.641	4.939x10^{-5}	7.696	227.9	9.648	1.919	3.7143
Jupiter	1898.	1.461x10^{-1}	7.696	778.6	9.648	1919	24.791
Saturn	568.4	4.374x10^{-2}	7.696	1433.5	9.648	1.919	10.44
Uranus	86.83	6.682x10^{-3}	7.696	2872.5	9.648	1.919	8.8709
Neptune	102.4	7.883x10^{-3}	7.696	4495.1	9.648	1.919	11.147
Pluto	0.012	9.619x10^{-7}	7.696	5870	9.648	1.919	0.5842

It is now of interest to involve the measurements of magnetic strength found by NASA flyby mission.[5] Equation (1-7 b) will be used to see if that equation can correctly predict the magnetic fields of planets and sun.

Determination of Relative permeability

However, as illustrated in Table 1-2, the prediction requires a relative permeability μ_r to explain the discrepancy between those calculated by equation (1-7 b) and the available experimental data of NASA.

Table 1-2 Various Calculation of Magnetic Data

Celestial body	B calculated using equation (7 b) (Tesla)	B_{NASA} data (Tesla)	B_{NASA}/B calculated = μ_r
Mercury	1.43280990×10^2	$3.00000000 \times 10^{-7}$	$2.09378788 \times 10^{-9}$
Venus	3.43368521×10^2	$3.00000000 \times 10^{-8}$	$8.73696864 \times 10^{-11}$
Earth	3.78968386×10^2	$3.05000000 \times 10^{-5}$	$8.04816475 \times 10^{-8}$
Mars	1.43305307×10^2	$5.00000000 \times 10^{-9}$	$3.48905431 \times 10^{-11}$
Jupiter	9.26729966×10^2	$4.20000000 \times 10^{-4}$	$4.53206452 \times 10^{-7}$
Saturn	4.03440827×10^2	$2.00000000 \times 10^{-5}$	$4.95735649 \times 10^{-8}$
Uranus	3.39864601×10^2	$2.30000000 \times 10^{-5}$	$6.76740088 \times 10^{-8}$
Neptune	4.28636484×10^2	$1.40000000 \times 10^{-5}$	$3.26617088 \times 10^{-8}$
Pluto	2.29172618×10^2	Not Measured	
Sun	1.06070435×10^4	$1.00000000 \times 10^{-4}$	$9.42769773 \times 10^{-9}$

The nature of this disagreement is not clear since the measured magnetic data by NASA's flyby mission are not directly measured at planet's surfaces but are rather modeled out from some distance away from the celestial body in question. Diamagnetism appears in all materials, and is the tendency of a material to oppose an applied magnetic field, and therefore, to be repelled by a magnetic field. However, in a material with paramagnetic properties (that is, with a tendency to enhance an external magnetic field), the paramagnetic behavior dominates.

Finding the Pole Strength

Thus, despite its universal occurrence, diamagnetic behavior is observed only in a purely diamagnetic material. Furthermore, in a diamagnetic material, there are no unpaired electrons, so the intrinsic electron magnetic moments cannot produce any bulk effect. In these cases, the magnetization arises from the electrons' orbital motions, which can be understood classically, as follows: When a material is put in a magnetic field, the electrons circling the nucleus will experience, in addition

---

[5] NASA's planetary fact sheet, Author: D R Williams, NSSDC, Mail CODE 690, NASA Goddard Space Flight Center, Greenbelt, MD 20771. USA.

to their Coulomb attraction to the nucleus, a Lorentz force from the magnetic field. Depending on which direction the electron is orbiting, this force may increase the centripetal force on the electrons, pulling them in towards the nucleus, or it may decrease the force, pulling them away from the nucleus. This effect systematically increases the orbital magnetic moments that were aligned opposite the field, and decreases the ones aligned parallel to the field (in accordance with Lenz's law). This results in a small bulk magnetic moment, with an opposite direction to the applied field. We appreciate this Lorentz force as being associated with the charges per unit mass Q and q or Θ which is the charge that enters in to the magnetic field descriptor.

$$g = \frac{c^2 \mu_o \theta^2 M}{4\pi r^2} = \frac{\mu_o P_1 P_2 M}{4\pi r^2} \qquad \text{1-8 a}$$

In above equation the variables P_1 and P_2 are the pole strength and defined by $P_1 = P_2 = c\Theta$. This is so because in most bar magnets their pole strengths are equal. We may now write the magnetic field of a planet:

$$B_{Plaanet} = \frac{c\mu_o \theta M}{4\pi r^2} = \frac{\mu_o PM}{4\pi r^2} \qquad \text{1-8 b c}$$

The equation (1-8 b c) gives definition of planetary magnetic field. In order to evaluate the relative permeability μ_r we require using the NASA data.

$$B_{NASA} = \frac{\mu_o \mu_r P_1 M}{4\pi r_2} \qquad \mu_r = \frac{4\pi B_{NASA} r^2}{\mu_o P_1} \qquad \text{1-9 a b}$$

Now since $B_{Planet} = \frac{\mu_o PM}{4\pi r^2}$ we can evaluate μ_r :

$$\mu_r = \frac{B_{NASA}}{B_{Planet}} \qquad \text{1-10}$$

Table 3 has a list of the magnetic fields measured by NASA, B_{NASA}, calculated magnetic field, B_{Planet}, using equation (1-8 b) and μ_r from equation (1-10). The pole strengths P_1 and P_2 are calculated as below. It is convenient to define the Newton's g-field and the Faraday B-field both in terms of charges and poles:

$$g = \frac{c^2 \mu_o \theta^2 M}{4\pi r^2} \qquad B_{Planet} = \frac{c\mu_{c\ \theta M}}{4\pi r^2} \qquad \text{1-11 a b}$$

$$g = \frac{\mu_o P^2 M}{4\pi r^2} \qquad B_{planet} = \frac{\mu_o PM}{4\pi r^2} \qquad B_{NASA} = \frac{\mu_o \mu_r PM}{4\pi r^2} \qquad \text{1-12 a b c}$$

When this is done the values of Θ is in In the case of the B field one can see that $(c\Theta M)$ has a distinctly different magnetic pole for each planet:

$$P = c\theta \; \frac{mc}{kg} \qquad \text{or} \qquad \frac{Am}{kg} \qquad \text{1-13 a}$$

6

This is a constant units of C/kg, and, *P* is in the units of ampere-meters per square root of mass for all planets.

Table 1-3 Magnetic Data of the Planets

Planet	Mass, (x10^{24}) kg	Radius, R of Planet (m)	B_{NASA} (Tesla) Ref.[1]	B_{Planet} (Tesla)	μ_r	Pole (Am/kg)
Mercury	0.3302	2.43950 x10^6	3.000 x10^{-7}	1.4331 x10^2	2.0938x10^{-9}	2.5834 x10^{-2}
Venus	4.8685	6.05200 x10^6	3.000 x10^{-8}	3.4361 x10^2	8.7370x10^{-11}	2.5834 x10^{-2}
Earth	5.9736	6.3780 x10^6	3.050 x10^{-5}	3.7952 x10^2	8.0481x10^{-8}	2.5834 x10^{-2}
Mars	0.64185	3.9360 x10^6	5.000 x10^{-9}	1.4372 x10^2	3.4891x10^{-11}	2.5834 x10^{-2}
Jupiter	1898.60	7.1492 x10^7	4.200 x10^{-4}	9.6034 x10^2	4.5321 x10^{-7}	2.5834 x10^{-2}
Saturn	568.46	6.0280 x10^7	2.000 x10^{-5}	4.0469 x10^2	4.9574 x10^{-8}	2.5834 x10^{-2}
Uranus	86.832	2.5559 x10^7	2.300 x10^{-5}	3.4326 x10^2	6.7674 x10^{-8}	2.5834 x10^{-2}
Neptune	102.43	2.4764 x10^7	1.400 x10^{-5}	4.2968 x10^2	3.2662x10^{-8}	2.5834 x10^{-2}
Pluto	0.0125	1950 x10^6		2.7859 x10^1	(estimated)	2.5834 x10^{-2}

Step Relationship between Relative Permeability & Gravitational Variables

In order to find a reason for the relative permeability's it was conjectured that the gravitational field of the sun and the planetary distance from the sun might both intervene with the NASA measurements. In Table 1-4 we have compiled, a natural logarithmic relation between relative permeability found for each planet upon division by its solar g and the distance between a planet and the sun that is:

$$D = \frac{-\ln(\frac{\mu_r}{kg})}{planetary\ g} = \frac{-\ln(\frac{\mu_r}{m})}{g} \qquad \text{1-13 b}$$

Graph 1 is plotted using values given in Table 1-4.

Table 1-4 Plotting Data

Celestial body	Distance between Planet and Sun, (meters)	D (Black dots calculated using equation (13 b)
Mercury	5.8344 x10^{10}	1853.45612
Venus	1.0846 x10^{11}	6984.99233
Earth	1.496 x10^{11}	7828.22697
Mars	2.2814 x10^{11}	28748.5844
Jupiter	9.2752 x10^{11}	349883.978
Saturn	1.4601 x10^{12}	1190991.67
Uranus	2.87082 x10^{12}	4686600.35
Neptune	4.49698 x10^{12}	11631467.2
Pluto	5.91369 x10^{12}	17145877.3

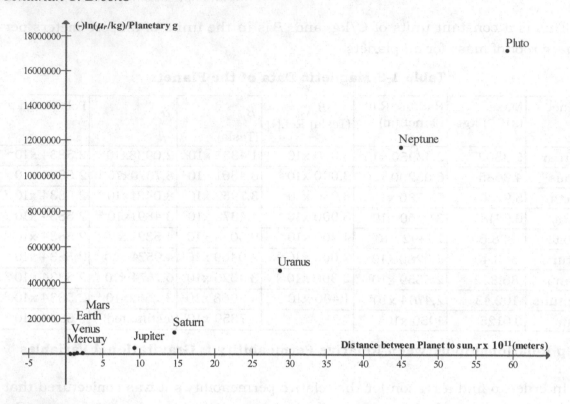

Planetary g versus Distance between Planet and Sun

Graph 1 has been plotted using data given in Table 4 and at first sight has an apparent parabolic nature; however, it is really a step function. In the present work, we have so far, mainly confined the calculations in terms of μ_r relative permeability. It is worth pointing out that the present analysis could have been analogously carried out in terms of relative permittivity (ε_r), since one could have written equation (7 b) as:

$$B_{Planet} = \frac{c\mu_o\theta M}{4\pi r^2} = \frac{\theta M}{c\varepsilon_o 4\pi r^2} \text{ and with } B_{NASA} = \frac{\theta M}{c\varepsilon_o\varepsilon_r 4\pi r^2} \text{ leading to } \varepsilon_r = \frac{B_{Planet}}{B_{NASA}}. \text{ This gives}$$

the value of $\varepsilon_r = \dfrac{1}{\mu_r}$.

Environmental Modification of Maxwellian Expression of Newton's Gravity

The modification of Newton's gravitational constant has been a subject of conjecture at various times. Usually it is thought to perhaps depend either on time or the age of the universe. The English scientists E. A. Milne, Paul A. M. Dirac, and others suggested that the proportionality constant G in Newton's equation for gravitational force might not be constant. Searches for such an anticipated small change in G have been made by studying accelerations of the Moon and the reflections of radar signals from Mercury, Venus, and Mars. Weinberg suggested that location might be a factor. The variation of *G* might indeed occur at different places. If, as a thought experiment, we consider the Cavendish[6] experiment with some electrons, as a negative parcel, added to one of the balls. From the increase of q, the electron density, one would expect to find an increased attraction between the two balls due to the increase in q/m of one ball. If the two were sufficiently close then one might see a touching; and then finally, a repulsion of the two spheres due to equilibration of charge. This has often been observed in the scattering of bits of paper from a Van de Graaf generator or the separation of two balsa balls after one is electrified. What has not been presented is an equation covering that first attraction wherein only sufficient electrons were added to perhaps increase the attraction and shorten the distance between two metallic balls so treated. If such an attraction as this occurs in outer space as an ejection/eruption of electrons during a supernova explosion, then there would be a capture and a concomitant greater attraction of mass that is implied by the use of the term dark matter. There is thought to be no general way to calculate the density of electrons in their usual orbit in ordinary matter. The Poisson is the equation that anticipates the electron density of matter. Dark or heavy matter can be explained by a required capture of electrons in outer space which will bolster local gravity and hence Newton's universal gravitational constant. An experimental determination of the gravitational strength of attraction of dark matter is probably the best way to determine Poisson electron density $\nabla \phi$ in space.

$$g = \frac{K\theta^2 \nabla^2 \phi}{r^2} \qquad \text{1-14}$$

This modification might well influence the expression of equation (15) of Einstein's general relativity.[7] It would likewise affect equation (16) of Hawking radiation formula for a blackbody.[8] Mir is the irreducible Schwarzschild mass. Einstein's field equation, independently arrived at by Hilbert is:

$$R_{u,v} - \frac{1}{2} g_{u,v} = 8GT_{u,v} \qquad (1\text{-}15)$$

$$M^2 = M_{ir} + \frac{s^2}{(4\pi GM_{ir})^2} \qquad (1\text{-}16)$$

Precession Related to Quantum Gravity:

---

[6] H. Cavendish, *Philosophical Transactions of the Royal Society of London*, (part II) 88 p.469-526 (21 June 1798).

[7] H. C. Ohanian and R Ruffini, <u>*Gravitation and Spacetime*</u>, W.W/ Norton & Company, NY and London, p. 382.

[8] S. W. Hawking and I. Werner, eds.: General Relativity, Cambridge University Press, Cambridge © 1971, p. 538ff.

Precession is found in all planets, particularly Mercury, since the precession of Mercury was one of the factors validating General Relativity. Methods using only Newtonian gravity have been used to verify precession as well. Equation (1-17) is given by Hawking and Israel for determining precession of planets solely by Newtonian methods.[9]

$$P = \frac{6\pi V^2}{(1-e)c^2} \qquad (1\text{-}17)$$

Compared to measured values reported on the internet we can use the graphical method used for relative permeability to obtain a parabolic relationship between the negative of the natural logarithm of precession P_w divided by planetary mass M and that divided by planetary gravity plotted against planetary distance r to the sun. The obviously parabolic fitted equation $-Ln\,(P_w/M)/$Planetary g = (4.2768E-19 r^2 - 364.11) has an average error of ca 5%. It is possible to obtain a linear step function for Table 1-5 below and the data of graph 1-1 by taking the logarithm base ten of the axes. Table 1-5 summarizes the result. In Graph 1-2 the summarized data is also plotted. The importance of Graphs 1-1 and Graph 1-1 and 1-3 is that both precession and relative permeability follow similar mathematics with respect to distance from central mass, the gravitational constant and mass of the orbiter. We have a non-integral quantum number with ties to both precession and relative permeability through distance and gravity. This is as close to an expression of quantum gravity as one can hope for.

There are several reasons to retain an interest in quantum gravity not the least of which is that it portends the "holy Grail" of physics. One would hope that the successes of quantum mechanics would be continued and that there would be a quantum world. The downside of that is the haunting reality that reality itself would be in a measure unattainable forever. There would then be a macro quantum. The next best thing might be a non-integral quantum number that could accept a large number of ordinary quanta.

Table 1-5 Data Pertaining to Graph 1-2

Planet	$-\ln(P_w/M)/$Planetary g (Plotted on Graph 2 as Black dots)	Factor x Fit	Fit to data = (4.2768 x10^{-19}R^2 - 364.11) (Plotted on Graph 2 as grey dots)	Precession P_w, sec/Year
Mercury	1343.84	1.23	1091.72	5.75
Venus	4974.39	1.07	4666.93	2.04
Earth	9207.43	1.00	9207.44	11.45
Mars	20400.31	.93	21895.72	16.28
Jupiter	394895.54	1.07	367566.15	6.55
Saturn	941699.68	1.03	911398.47	19.5

---

[9] H. C. Ohanian and R Ruffini, op. cit., p. 405.

Uranus	3633322.68	1.03	3524416.40	3.34
Neptune	9279170.08	1.07	8648520.06	.36
		Average = 1.05		

In equation (1-1) making use of the minus sign will send us into the atomic realm. The LHS equated to *G* can be shown to proclaim quantum gravity equation is equivalent to the de Broglie hypothesis in the atomic realm. For the solar realm the value of quantum number n, is neither an integral number nor is it well defined. The name for the problem of reconciling Quantum Mechanics with General Relativity is Quantum Gravity.

Non-integral Quantum Numbers via a macroscopic Planck's Constant

A quantum number can be calculated directly by using equation (1 c) or by modification to (18 a). We have obtained the set of quantum numbers *n* as {2.1, 2.9, 3.4, 4.2, 8.5, 10.6, 14.9, 18.7, and 21.4} for all planets using equation (1 c). If we add masses M, m and radii to equation (1 c), i.e. to $\frac{nhV}{qmr} = \frac{GM}{r}$, we can recover the de Broglie hypothesis for atomic phenomena.

$$\frac{n\hbar v}{qmr} = rg \qquad\qquad \text{1-18 a}$$

$$\frac{n\hbar}{q} = mvr \qquad\qquad \text{1-18 b}$$

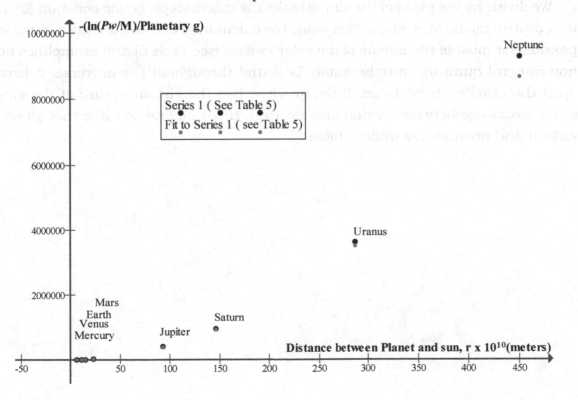

11

Planetary g versus Distance (*r*) between planet and sun.

The solar quantum number *n* is not an integer. To insure small numbers starting with near unity we can define a macroscopic action constant η for equation (1-18 a, and 1-18 b). Thus the solar quantum number can kept small by being defined by the charge on an electron and a unit kilogram of central mass. The chief variables are the radius *R* and orbital velocity *V*. The factor containing the unit of natural logarithms is fortuitous. The factor η in equations ((1-19 a) to (1-19 c)) is for the real quantum number for our solar system.

$$\frac{\eta\, hV}{q} = G$$

1-19 a

The above equation was the LHS of the Rydberg Rearrangement. When Planck's constant is increased by the Avogadro number for the kilogram and by the ratio of planetary speed to the speed of light we can obtain low numbered values for ten planets. Note the charge, q, is multiplied by velocity now.

$$\eta = \frac{G}{h\left(Vc\,q\right)N_{avogadro}}$$

1-19 b

The resulting non-integral solar planetary quantum numbers are {0.704, 0.821, 1.013, 0.292, 2.535, 3.595, 4.501, 5.143, and 6.027}. These are still not integers.

We divide by the mass of the sun to make the macroscopic action constant for any unit central mass, M, system. This made the calculation of small quantum numbers possible for most of the moons of the solar system (see Table 6) and exemplifies how non-integral quantum numbers may be found throughout the universe. It leaves open the graviton hypothesis. If the graviton has the virtual nature of the gluon of the exchange between proton and electron, then the graviton like that gluon is virtual and presumably undetectable.

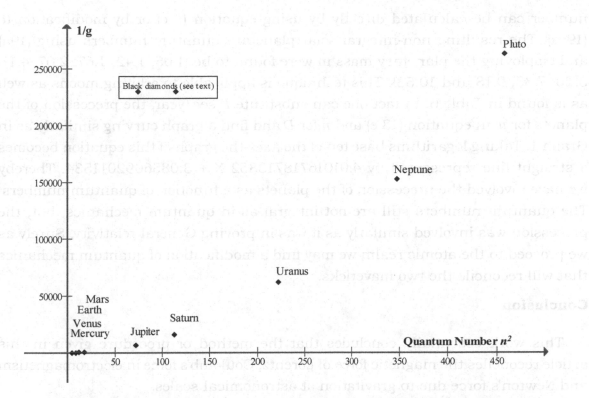

Relationship of Quantum Number n^2 to inverse of planetary gravity 1/g

It was found that Graph 1 seemingly showing a parabolic nature is not a parabola. By taking the logarithm base ten of both axes an equation was found with requisite accuracy. Allowing the logarithm base ten of distance to be the X value: Y = 2.00354512594032 X – 18.4119361085405. The standard deviation of the slope is 0.10040019474421 and the average deviation out is 2.8 x10^{-5}σ. The standard deviation of the intercept is 0.0237000774093292 and the average out is 0.03569. Using this format one can easily arrive at an estimate of the $\left(\frac{-\ln(\mu_o)}{m}\right)/g$ values of the *B* Field that NASA might have measured for the moons of Jupiter, Saturn, Uranus and Neptune. The results can be found in Table 6 all giving curves similar to that of Graph 1. One may plot the $\left(\frac{-\ln(\mu_o)}{m}\right)/g$ value against either n or the moon to planetary distances and find smooth curve. The disparities in central mass values do not permit all to be shown on the same plot because central masses of the moons are different and the μr values indicated are much too varied to give practical plots.

Using the slope and intercept method reasonable values of the $\left(\frac{-Ln(\mu_o)}{mg}\right)$ can be secured for all but the retrograde moons of Uranus. Values at these high quantum numbers approach zero essentially indicating that there would be no apparent measureable magnetic field. The data of Table 6 is included only as an estimate of the range of variability of the relative permeabilities of the planetary moons. The data could be presented as a function of the distance of the moons from their planets but using the definition of the macroscopic action, η and hence we have a good illustration of one of the features of quantum gravity. As indicated earlier a quantum number can be calculated directly by using equation (1 c) or by modification to (19 a). The resulting non-integral solar planetary quantum numbers using (19a) and employing the planetary mass m were found to be {1.05, 1.42, 1.67, 2.07, 4.17, 5.23, 7.33, 9.18 and 10.53}. This technique is applicable to orbiting moons as well as is found in Table 6. In fact one can substitute P_w sec/year, the precession of the planets for μ_r in equation (13 e) and n for D and find a graph curving similarly as in Graph 1. Taking logarithms base ten of the axes the graph of this equation becomes a straight line represented by 4.01016718713852 X + 3.0856092011534. Thereby we have involved the precession of the planets as a function of quantum numbers. The quantum numbers still are not integral as in quantum mechanics; but, the precession was involved similarly as it was in proving General relativity. Surely as we proceed to the atomic realm we may find a modification of quantum mechanics that will reconcile the two mavericks.

Conclusion

This work essentially concludes that the method or procedure given in this article reconciles the magnetic force of Lorentz, Coulomb's force in electromagnetism and Newton's force due to gravitation at astronomical scales.

Table 1-6: Quantum Numbers, Relative Permeability's & Magnetic fields of Moons.

Planet	Moons	N	$-\ln\left(\frac{\mu_r}{m}\right)/g$	μ_r	$B = c\mu_r\Theta M/4\pi r^2$ (Tesla)	Predicted B Field (Tesla)
Jupiter						
	Io	2.86	1.109 x10^2	4.001 x10^{-12}	7.025 x10^1	2.811 x10^{-10}
	Europa	3.61	1.935 x10^2	1.019 x10^{-1}	6.330 x10^2	6.449 x10^1
	Ganymede	4.56	4.462 x10^2	5.293 x10^1	6.934 x10^2	3.670 x10^4
	Callisto	6.03	1.485 x10^3	4.298 x10^{-1}	6.012 x10^2	2.584 x10^2
Saturn						
	Pan	3.01	7.931 x10^{-3}	4.870 x10^{15}	6.571 x10^{-10}	3.200 x10^6
	Atlas	2.99	7.724 x10^{-3}	5.967x10^{17}	8.268 x10^{-8}	4.933 x10^{10}
	Prometheus	3.01	7.917 x10^{-3}	1.571x10^{17}	2.123 x10^{-8}	3.334 x10^9
	Pandora	3.04	8.218 x10^{-3}	2.166 x10^{17}	2.821 x10^{-8}	6.111 x10^9
	Epimetheus	3.14	9.344 x10^{-3}	5.185 x10^{17}	5.940 x10^{-8}	3.080 x10^{10}
	Janus	3.14	9.344 x10^{-3}	1.979 x10^{18}	2.267 x10^{-7}	4.488 x10^{11}
	Mimas	3.48	1.411 x10^{-3}	3.742 x10^{19}	2.840 x10^{-6}	1.063 x10^{14}
	Enceladus	3.93	2.313 x10^{-2}	8.271 x10^{19}	3.834 x10^{-6}	3.171 x10^{14}
	Tethys	4.38	3.556 x10^{-2}	6.080 x10^{20}	1.835 x10^{-5}	1.115 x10^{16}
	Telesto	4.38	3.556 x10^{-2}	7.805 x10^{15}	2.355 x10^{-10}	1.838 x10^6
	Calypso	4.38	3.556 x10^{-2}	4.933 x10^{15}	1.489 x10^{-10}	7.343 x10^5
	Dione	4.95	5.813 x10^{-2}	1.034 x10^{21}	1.910 x10^{-5}	1.975 x10^{16}

Planet	Moons	N	$-\ln\left(\frac{\mu_r}{m}\right)/g$	μ_r	$B = c\mu_r\Theta M/4\pi r^2$ (Tesla)	Predicted B Field (Tesla)
	Helene	4.95	5.813×10^{-2}	5.084×10^{6}	9.394×10^{-20}	4.776×10^{-13}
	Rhea	5.85	1.137×10^{-1}	2.452×10^{21}	2.318×10^{-5}	5.683×10^{16}
	Titan	8.91	6.133×10^{-1}	1.329×10^{23}	2.338×10^{-4}	3.107×10^{19}
	Hyperion	9.71	8.652×10^{-1}	1.743×10^{19}	2.174×10^{-8}	3.788×10^{11}
	Iapetus	15.21	5.228	1.851×10^{21}	3.834×10^{-7}	7.095×10^{14}
	Pheobe	29.01	6.947×10^{1}	9.844×10^{17}	1.541×10^{-11}	1.517×10^{7}
Uranus						
Prograde	Cordelia	4.60	9.972×10^{7}	4.914×10^{-9}	2.844×10^{-1}	1.398×10^{-7}
	Ophelia	4.78	1.078×10^{8}	1.048×10^{-9}	3.108×10^{-1}	3.256×10^{-10}
	Bianca	5.02	1.185×10^{8}	1.325×10^{-10}	3.263×10^{-1}	4.323×10^{-11}
	Cressida	5.13	1.238×10^{8}	4.858×10^{-11}	2.747	1.335×10^{-10}
	Desdemona	5.16	$1.256E \times 10^{8}$	3.437×10^{-11}	4.682×10^{-1}	1.609×10^{-11}
	Juliet	5.23	1.289×10^{8}	1.809×10^{-11}	6.642×10^{-1}	1.201×10^{-11}
	Portia	5.30	1.324×10^{8}	9.268×10^{-12}	9.506×10^{-1}	8.810×10^{-12}
	Rosalind	5.46	1.401×10^{8}	2.110×10^{-12}	5.012×10^{-1}	1.057×10^{-12}
	Cupid	5.64	1.499×10^{8}	3.258×10^{-13}	6.823×10^{-2}	2.223×10^{-14}
	Belinda	5.66	1.508×10^{8}	2.730×10^{-13}	7.919×10^{-1}	2.162×10^{-13}
	Puck	6.05	1.723×10^{8}	4.381×10^{-15}	1.143	5.007×10^{-15}
	Mab	6.45	1.957×10^{8}	4.898×10^{-17}	1.010×10^{-1}	4.947×10^{-18}
	Miranda	7.42	2.592×10^{8}	1.654×10^{-2}	3.065	5.069×10^{-2}
	Ariel	9.02	3.827×10^{8}	1.748×10^{-11}	1.044×10^{1}	1.825×10^{-10}
	Umbriel	10.58	5.269×10^{8}	1.458×10^{-23}	8.864	1.293×10^{-22}
	Titania 1/8	13.62	8.734×10^{8}	5.981×10^{-52}	1.467×10^{1}	8.775×10^{-51}
	Oberon	15.76	1.169×10^{9}	1.120×10^{-76}	1.343×10^{-3}	1.504×10^{-79}
	Francisco	42.66	2.134×10^{-9}	0.000		
	Caliban	55.47	1.449×10^{10}	0.000		
	Stephano	58.35	1.603×10^{10}	0.000		
	Trinculo	60.39	1.717×10^{10}	0.000		
	Sycorax	71.99	2.440×10^{10}	0.000		
	Prospero	83.58	3.289×10^{10}	0.000		
	Setebos	86.19	3.498×10^{10}	0.000		
	Ferdinand	94.31	4.188×10^{10}	0.000		
Neptune	Nereiid	44.59	1.255×10^{1}	9.972×10^{-1}	2.684	2.677
	Naiad	4.17	9.443×10^{-4}	9.972×10^{-1}	2.460×10^{-1}	2.453×10^{-1}
	Thalassa	4.25	1.018×10^{-3}	9.972×10^{-1}	2.586×10^{-1}	2.579×10^{-1}
	Despina	4.35	1.121×10^{-3}	9.972×10^{-1}	3.778	3.767
	Galatea	4.73	1.560×10^{-3}	9.972×10^{-1}	6.629	6.611
	Larissa	5.15	2.200×10^{-3}	9.972×10^{-1}	1.388	1.384
	Proteus	6.51	5.637×10^{-3}	9.972×10^{-1}	7.675	7.654
	Triton	11.31	5.146×10^{-2}	9.972×10^{-1}	3.023×10^{1}	3.014×10^{1}

The calculated local and planetary gravity are in exact agreement with those deduced on the basis of Newton's original formulism. That gravity was an electromagnetic phenomenon was predicted by Einstein who worked in his latter part of life to prove it. The deduced relative permittivity of each planetary body, as measured by NASA flyby missions, can be mathematically accounted for in terms of their distance from the sun and its gravitational field. While one might claim that we have only presented only a mathematical artifact we counter with the original impetus obtained from the rearranged Rydberg expression. It expresses a mathematical equation that implies that Newtonian gravity is an electromagnetic phenomenon. General relativity is the generalization of special relativity to include gravitation. It is worth noting that the representation of Newton's G value using Coulomb's law constant in areas where there is an overabundance of electrons would alter the constant Θ due to a change in the electron density. We thereby have a candidate for dark matter and a modification of general relativity. If gravitons are virtual gravitons then the fractionation of Planck's constant by non-integral quantum numbers should be acceptable. Certainly the pair of step functional relations of non-integral quantum number to gravity and to precession constitutes a step towards quantum gravity since the macroscopic equivalent of Planck's constant is a derived value. Although we have demonstrated a defensible quantum-like gravity, the reconciliation between quantum mechanics and general relativity still remains to be proven.

In this work, we have evaluated the planetary dynamics of Newton in order to reconcile it's unification with Coulomb's law of force in electrostatics and Lorentz force in magnetism. Magnetic fields of the planets and the sun have been calculated. We show here, explicitly, that magnetic and electric fields of our solar system are in exact mathematical accordance with Newton's law and available NASA's flyby data [1]

Chapter Two

Forming the Atomic and Solar Field Connection

In this chapter I intend to expand the statement that one obtains the unified field of the atom from the Rydberg expression and in the following chapter I will present numerical examples of radial calculations and spectral matches for hydrogen and helium. The radial description of the unified field is a phenomenon that exemplifies the position of electrons, their orbits, and velocities in atomic orbitals. At the outset I am appealing to centrifugal force to correlate the energy and radii of hydrogen and helium instead of the Lagrangian or Hamiltonian. This will have the effect of eliminating the objections raised as to the symmetry and gauge breaking that occurs when the Schrödinger equation is transformed into General Relativity. It may not entirely supplant quantum mechanics particularly in calculations outside the atom; but, abandoning the Schrödinger equation for the unified field expression may well allow the reconciliation of atomic phenomena with General Relativity because the gauge transfers simply from the unit mass of the electron to the unit mass of the kilogram. One of my objections to the Schrödinger equation is that the inclusion of Planck's constant implies that angular momentum is conserved, a view that seems to be prevalent but one that the unified field cannot support.

On would not expect from equation 2-1 a below that a radius of hydrogen could be extracted:

$$(N+5)F + 2F' - (N-28)N^{-1} + 7 = 0 \qquad \text{2-1 a}$$

The function F is defined as $(1-\alpha)^{-1}$. During the course of the following chapters I will call it the Bessel function, a Laplacian, a Poisson and the d'Alembertian. It is the latter because I realized that the last two terms being zero reflect the zero nature of time in the time dependency direction. Time does not generally have a direction except in the spacetime of general relativity. The first two terms of equation 2-1 a can be called perpendicular vectors because the differentiation of the first with respect to a variable α yields the second. Obviously, I chose the latter two zero terms orthogonal to the other two. With time being zero I have a zero Laplacian as a vector in the time direction; but, if I find α as a function of N then I have a Poisson which is a scalar density function. Together they constitute the d' Alembertian.

The Rydberg rearrangement provided the unified field and the ad hoc inclusion of charge density satisfies Maxwell's laws. Taken together they dispel any thoughts of dark matter as a new kind of matter. Although this definitely challenges the Standard Model, it has the potential to embellish physics discipline in all orbiting systems where the unified field has long been awaited. It also frees us from some of the internal difficulties of the Standard Model, namely the infinities. These procedures will include not only atoms, but also will also involve the proton, the neutron and all other sub atomic orbiters. The account of my work that follows will often cover the same ideas over and over again hopefully with some embellishment which I hope that the reader will discern

One can find many assertions made in former times that were then the best available and most useful at that moment. Because many avenues were discarded and others embellished often times incorrectly, it is not necessary to travel every road and encompass every history to demonstrate progress. One of the tenets of quantum mechanics is the notion of the quantum itself. The original quantum was mistakenly connected with the variable n that can be described as a method for counting the entering photons that excite the electron into a more energetic orbit. The quantum is more correctly connected to an amount of energy. Einstein is credited with the enunciation and characterization of the quantum. The quantum as employed by quantum mechanics may fade with the grin of the Cheshire cat; although a variable n will *reappear as integrally different real integers on the number line.* As we go to the atoms following helium, and a *real numbered quantum* attribute will surface in relation to precession as we saw in the first chapter. This of course requires calculation of the macroscopic analog of Planck's constant. So often now the quantum is the just number n of the loops in a wave fastened in a box of a given dimension. This description of the quantum bears latterly no numerical relation to the quantum defined by Einstein. No wonder he saw quantum mechanics as incomplete. It is also no wonder that when the Hamiltonian of a particle in a box acquires an extra dimension, the product of time and the speed of light, the wave cannot find a home. This presents an even more difficult picture than for three dimensions should a fourth dimension of general relativity be added. Since the Lagrangian was used in formulating general relativity one might conclude that there would be no difficulties of compatibility between different versions of quantum mechanics and general relativity; but, the Lagrangian of the latter does not include Planck's constant. In regard to non-cosmic applications of quantum attributes, Planck's constant appears only as a multiplier of the frequency of the photon. A large part of the standard theory depends on the Schrödinger wave either as a Hamiltonian or a Lagrangian. Curiously it is usually stated that any solution to the Hamiltonian can be added to another with the result being also a solution.

When assessing the photon's entrance into the atom it makes no difference if the high energy spectrum or the low energy of the spectrum is inserted first into the calculation to find a radial difference in magnitude. One wonders, would it not be possible to start with any photon and hit an electron in any state? Mathematically one could even make up a spectral component or two and insert them anywhere amidst the published spectra. However, given the acknowledged spectra and the dictum that says the atom only accepts certain definite jumps it could be said that any photon entering ensemble should be welcomed by any member. Its emission will merely augment the display. It would seem that any ensemble could find a mode in which conventional spectra could be forthcoming.

I find it questionable to consider the kinetic and potential energy expressed as the sum or difference of these components as relevant to circular motion when centrifugal theory of motion seems to be sufficient. After all the kinetic energy is half the potential energy in Newtonian systems. The real irrationality was the expression of radii as multiples of the Bohr radius. For every elliptical orbit there is a circular orbit with the same frequency. It makes no difference which one is found first. Given the circular orbit it seems that the electron in that orbit will perceive a different electronic environment as it orbits. This means a different electron density or a variable Poisson. Manipulating the calculation of the Poisson will be how I can deal with elliptical orbits.

At present we shall provide a brief history and critique of the more salient serendipities and missed-serendipities of what I call *attribute mechanics*. One of the more advantageous solutions in verifying my orb calculations was my finding a numerical solution to the Associated Legendre equation which is the radial portion of Schrödinger's wave equation. This applied to an atom receiving energy from a photon and obviated Planck's constant other than as a descriptor of the energy of the photon. Simply put it was another source of a portion of an equation that I derived for another purpose. This was the steady state d' Alembertian that first yielded the digits of Bohr radius as I examined relations of the variables with a slide rule. Computers had not yet become available in the sixties. Upon further development I found the algorithm fitting the radii of the atoms of the periodic table using a matrix of *real variables* of the number line. I allowed *five integral numbers* between s orbitals. This allowed for s, p, d, f, and g orbitals which ensemble can avoid the weird picture one sees accorded to the "f orbital". I shall explain that avoidance later; but it is easy to see that extra space is needed. Thus, these matrices of real variables defined the number line upon which a modified integral quantum number, n, floated. I gave the name *attributes* to elements of the orbital matrices of real numbers whose integral differences were arguments of the parametric equations. The summation of these matrices of r values could be iterated on a spreadsheet to yield the known radii of atoms by varying the initial value of N. Later the subsequent g field, velocity, Poisson, and energy values or averages needed could all be calculated from that given N which corresponded to any orbital of a given element. All atoms of the table were able to be similarly constituted. Each element was found to have a *principal attribute, N1s,* indicative of its 1s orbital. The group of the attributes of the 1s orbitals was found to be periodic with the atomic number; but, inversely periodic with relation to the normal periodicity of radii with atomic number which is normally presented. The cornerstone, or real 1s matrix element thus I increased by five in going from *1s* to *2s* to *3s* and on. What normally would be designated a *l* quantum numbers were the 5, 6, 7 or 10, 11, 12 added to the 1s value so that one could define a matrix by a principal attribute to which a continuous unit value was added and was applicable as long as that position was filled by the usual aufbau principles. This allowed for a g orbital mostly never to be filled except as for the super-heavy elements envisaged by Glen Seaborg. Since the Bohr hydrogen radius fell out at an attribute of ca. 40.54 and this being a half filled orbital I determined that should two electrons be in the orbital as determined by aufbau principles that I would have to double the value of the single radius at that attribute value for the contribution of the pair. The p, d, and f orbitals were prorated for occupancy.

Later when I returned to the inclusion of spectra of hydrogen I independently developed another new idea, the unified field, based on adding gravitational and electric fields. There is no kilogram test particle in the atom. Instead the electron mass of unity is used. From the spreadsheet matrix values of N and with the unified field the Aufbau velocities could be found. This enabled the spectral energies of hydrogen and helium to yield their change in radius with energy received since the 1s electron(s) moved as a unit. Both the electric and gravitational field are required to have as their basis the unit electron mass. I used the unit electron mass as test particle with added gravimetric and electromagnetic contributions. Previously everyone had concluded that if this were done using a test particle of a kilogram that gravity was negligible. Whoever conceived of a test particle the size of a kilogram in the atom? Soon thereafter I derived this same unified field expression from a rearrangement of the Rydberg equation. In the process of that derivation, a square root was taken. The positive and negative of that root allowed a dual interpretation. From this I was able to deduce that not only that there was a unified field of the atom but there was also an equivalence of gravity and the electric field that could only apply to the solar system. Later I conditioned the entire unified field by the Poisson, available from the Bessel function. I realized that Newton had had no need of a Poisson because it was unity; however, in the area of general relativity where Einstein employed Newton's constant Coulomb's law constant could now be employed. This will require a Poisson correction for bursts of supernovas that might spew electrons indiscriminately landing on orbiting masses. This would increase the apparent value of Newton's constant which is being interpreted as an atypical matter termed "dark" matter. Today what is termed dark energy effect may also be attributed to certain other placements of ejected electrons. One reason for not supposing an electromagnetic version of gravity was that the former phenomenon was bidirectional. Chapter one dispelled that view. There is a continuation of some form of gravity down to sub-atomic dimensions as well as the attraction of their electric charges. This differs from e^2/r^2 usually applied. Recall that Newton encapsulated *Kepler's law constant into his Universal Gravitational* by multiplication of the universal constant of gravity by the central mass. I found it to be a fact of nature that allows me to similarly connect *Coulomb's law constant with Newton's universal gravitational constant*. I am able now not only claim a classical atom with a unified field but also to assert the proof that *gravity and electromagnetism* are one in the solar system as I discussed in the first chapter. The later phenomenon is an idea that Einstein and others conjectured yet did not show proof. It is also clear that the gravitational portion of the atomic case has an electromagnetic exposition which gives a far different picture than the Schrödinger equation suggests. The inclusion of the electron density, the Poisson, now obviates one difficulty that multiple orbits presented, namely that no method of handling the repulsive effects of electrons in atomic orbit was known.

To further recapitulate preceding events, in the early 60's I had occasion to derive an equation for the separation of two closely identical chemical compounds in a non-aqueous material. In those times we had only slide rules to do our computations. After communicating my result to my employers I continued to evaluate the equation. While fiddling with numbers the value 529 came under the line of the slide rule. I was able to iterate an algorithm which was undoubtedly accurate to over 8 places for the 1s orbital of hydrogen, the Bohr radius. As programmable hand held calculators became available I found a rule for adding the contribution of successive orbitals until I had, by iteration, a new 1s orbital (*principal attribute*) along with the requisite successive orbitals for iterating the radius of any given element. The Aufbau Principle governed devising each real numbered matrix. A hill climb algorithm which varied the 1s attribute sufficed to allow me to iterate the known radius of any given element when spreadsheet programs became available. Then I had an algorithm that was applicable throughout the table. Previously my first attempts had accorded a different rule to each area of the table.

In the 80's my Advanced Chemistry class needed a project. I guessed what I later derived from the Rydberg Rearrangement in equation 2-1. I just decided that the unified field had to be an additive function and this suggested what is in equation 2-2; but it was without the Poisson.

$$\frac{nhV}{2\pi qMm} = \frac{G}{q} = \frac{KQ\sqrt{\dfrac{mvr}{\hbar}}}{Mm} \qquad \text{2-1 b}$$

The radical evaluates as plus or minus and allows definition of two fields. The unit coulomb cancels (since it appears in K), taking the negative value of the radical and upon ad hoc addition of the Poisson (to both fields because of their electromagnetic equivalency) we have the unified field:

$$g = \frac{GM\nabla^2\phi}{r^2} + \frac{KQ\nabla^2\phi}{m\,r^2}$$

2-2 a

I communicated this to P.A.M. Dirac, who then at the University of Florida, but received no reply. At least Glenn Seaborg had given me a reply regarding my *earliest* attempts to calculate elemental radii. His response was negative so I searched for a better way. The salient feature of equation 2-1 was that the masses had to be in electron masses, {1836.15 for central mass and 1 for test mass}. My three advanced chemistry students and I each measured separately and then averaged values of the deuterium spectrum. However more was needed.

$$\alpha = \frac{(N-35)}{(N-28)} \; ; \; r = \frac{\alpha \ln(\alpha)}{\sqrt{N+6}}$$

2-3 a

They employed the parametric equations 2-3 and a computer to iterate the known radius for deuterium. Next one can iterate the N values that elicit the changes in radii associated with the $m_e v^2$ values of energy of successively entering spectral photons by using equations 2-2 b, c and 2-3 a in a spreadsheet. I will explain this and improve on the technique further in chapter three. These radii due to added energies are really changes in radius which I realized later must be taken from the original deuterium radius to determine the normal spiraling in of higher energy orbits. I tried for several years to justify equation 2-1 by dimensional analysis. Eventually, I found the reason in the derivation of equation 2.2 a, which stems from a rearrangement of the Rydberg equation.

I rearranged the Rydberg constant derived by Neils Bohr and achieved equation 2-1 wholly by algebraic means because I did not include the second reciprocal n² of the original Rydberg equation. The two sides could be divided by central and test mass, a perfectly allowable condition. This resulted in the equivalency of the units of Newton's Universal gravitational constant and those of Coulombs law when further divided by radii. But, I had a method of finding the radii to divide using the parametric equations. I also found the Poisson to condition the combination. Still one problem remained. Why should the mass, now in electron mass units give the correct answer? One might think that the Newton in electron mass units was surely numerically different than if it were based on the kilogram. The answer is that Kepler's law is given by the product of Newton's universal gravitational constant, G, and central mass, M. In Kepler's law the mass M is central mass while the orbiting mass is unity and the orbiting test mass of unity is hidden in one of the two kilograms in the denominator of Newton's law. One of the two kilograms in the denominator of Newton's law cancels the kilogram of its numerator leaving the other, now an electron mass, to cancel the central electron mass value of a proton, namely, 1836.15. Actually under gravity one gets this value if one divides the mass of the proton in kilograms by the electron mass in kilograms. Had P.A.M .Dirac concentrated on fields not forces he certainly would have realized this. It seems to me that no bar to inclusion of classical concepts in atomic calculation is left to question, not even the idea of having Coulomb's constant in a unified field itself. There is neither an orbiting coulomb nor an orbiting kilogram in any atom. It is not reasonable to compare the force per kilogram to the force per coulomb and conclude anything about the relative strength of the electric field compared to the gravitational field.

Now it is apparent that it is the ratio of the two *fields not forces* that is important. We can assert that here is a unified field and equation 2-1 describes it. Well, what happened to the extra negative charge on the electron? I removed it because so doing is essentially equivalent to equating it to unity as is seen in equation 2.2. It is customary when dividing a force by the orbiting mass or the charge on the electron to relegate the force to a field based on unit mass or unit coulomb. When the charge on the electron disappears it really is the charge per electron mass which is restored by dividing by Mm. Actually Coulomb's law constant was fashioned by analogy the Newton's universal gravitational constant. You can see that dividing by Mm requires an electron mass, m, to form in order to equate each of the end terms to the units of G. This removal is done when the *field F/q is calculated for the purely electromagnetic* atomic field; but, the field is still quoted as Newton's per (unit) coulomb. This coulomb unit cancels its coulomb counterpart in the Coulomb's law Constant.

Note what happens when you add the electric and gravitational fields.

$$g\frac{m}{\sec} = \frac{F}{em} = \frac{GM}{r^2}\frac{\nabla\phi}{}\frac{kg\,em}{kg\,em} + \frac{KQ}{r^2 m}\frac{\nabla\phi}{}\frac{kg\,em}{kg\,em} \qquad \text{2-3 b}$$

I only recently realized that because of the nature of the solar field that the Poisson should multiply both terms of the *unified field*. In the term involving Coulomb's law we see one mass unit apparently not canceling as we should expect. It hybridizes Coulombs law constant. If we consider the kg-kg pair cancelling in the Newton's law term we might as well have electron mass-electron mass units which employs a unit causing the Newton inherent in the constant K to be the same comparable value as it is in the term containing G. In other words the constants G and K are just required to be on a unit basis of the test charge and their usual value is the same whenever unit masses are employed in the denominator. A Newton whose units are electron mass m/sec² referenced to an electron mass is numerically the same as a Newton whose units are kg m/sec² referenced to a kg so that G and K have the same value as in ordinary practice. For years there has been controversy in gravitational theory as how to interpret, find, or manage the density function as is done in electromagnetic theory. It apparently is unity in the solar system; but, takes on the density calculated from the Bessel Function in cylindrical coordinates for atomic phenomena. As was mentioned in chapter one, unknown electron densities are very probably the reason for the phenomenon of dark matter. In certain other areas of the sky another placement of electrons that does not fall on orbiting matter may well induce some galaxies to appear to move outward from earth thus characterizing the phenomenon of dark energy.

At one early time in the development of quantum mechanics the key players were in their twenties for the most part with the possible exception of Planck, Einstein and Bohr. I am certain that at that time the major obstacle to classical expression during those years was twofold. There was the lack of knowledge of the required nature of a unified field; although, I believe that Bohr probably was close to it because he derived the Rydberg constant. Secondly, if Bohr was aware that Newton's law could be involved he very likely ignored it because no radial solution was known at that time. Louis de Broglie also possibly *could* have been aware of the Rydberg Rearrangement because he came up with something which can be found easily from $n\,\hbar\,v/\mathrm{Mm} = G$, one of the equations stemming from rearranging the Rydberg Equation.

It became clear that if equation (2-2 a) were true then there was another reason to be found.

$$\frac{nhv}{2\pi mr^2} = \frac{GM}{r^2} = g \qquad\qquad 2\text{-}4$$

$$\frac{n\hbar v}{mr} = rg = v^2 \qquad\qquad 2\text{-}5$$

From equation 2.5 it is a small step to derive $n\hbar = mrv$, the de Broglie relation. But, again, not having means for finding radii he took another direction which was questionable. If there must be a role for the de Broglie relation and it has to be in solar quantum gravity for which the quantum number is not integral. He received the Nobel Prize for predicting the dual nature of matter which was later proven by diffraction of electrons. *However, the more crucial lack in atomic theory was having no equation yielding radius or velocity.* This was the case because no solution to the radial Schrödinger equation was known at that time. Indeed a third unforeseen problem was having no method to appraise the Poisson electron density. The concept of a Poisson density for an electromagnetic field was known; but, not in the context of a radial solution to atoms. Those who tried to apply electronic repulsion found it a very difficult task. The fact of a density for the gravitational filed was but a conjectured question for such a density was neither understood in those times nor generally appreciated now. It is a requirement in Maxwell's Laws. The electron density in certain areas of outer space will change the effective value of Newton's constant G and possibly be the cause of what is described as dark matter.

Planck had solved the riddle of the ultraviolet catastrophe and Einstein had formulated the particulate notion of light called the photon. Bohr tried his best to formulate an atomic theory without using the notion of the photon. Schrödinger did not concur nor approve of Bohr's idea of quantum electron jumps. He envisaged the electron as being purely a *wave without orbit*. He and de Broglie were the proponents of electrons as waves. This wave became even more nebulous when applied to atoms above hydrogen where it was undoubtedly a composite. Also, Schrödinger's equation could not be solved exactly for hydrogen and atoms beyond hydrogen. It was popular to say that even if a velocity were measured there was no clue as to the whereabouts of the electron. *Any measurements* had to be external to the atom. Schrödinger's solutions were in phase space. I can show in ordinary space that the Poisson density stems from the separation of the radial solution Schrödinger's equation. In the current era it is said that graduate students spend most of their time in mathematics in learning how to solve the Schrödinger equation. The saving of time by using *attribute mechanics* would appear to be a reality.

An electron upon receipt of a photon generally should attain a tighter orbit. It took some time for me to interpret and calculate the spectra in this fashion. I had noted that the presence of orbitals *added* to the totality of the radii of the table and I assumed that there would be a similar behavior for spectra. However this is contrary to what is found in spacecraft orbiting the earth and so it should be and is for the orbits of hydrogen and helium accepting spectral energy.

Much is made of the uncertainty principle. It is connected to the idea of complementarity and the value of one of a pair of complementary variables in regard to a measurement of the other, ostensibly because the act of measurement, disturbs the system. I really wonder of what use knowing the electron's exact position can be, especially since it must be somewhere in one atom of an ensemble of many atoms! Why even enunciate the "uncertainty principle"? It must have been formulated from measurements external to the atom. True, the electron must be somewhere because of the nature of the ensemble. Also, because should we specify a circular orbit there is uncertainty as to where the electron would be if that the orbit were elliptical. I have calculated elliptical orbits by taking derivatives of the Poisson and using the variation of velocity to calculate the radial extreme from which eccentricities can be found. We shall not need to worry about uncertainty as a result of disturbing a system under measurement because we are calculating everything and not measuring anything. It occurred to me that because I had ignored a third derivative in relating the Bessel equation to Schrödinger's radial equation that I could recover it by differentiating the quadratic of the Poisson and let the Poisson cycle above and below these limits. The Poisson affects the basal circular radius and also the velocity. If the velocity were to vary using the conservation of momentum there is a variation of radius from normal that can be calculated. We now have an elliptical orbit. There is room for the electron to vary ±½ n from the attribute N of orbit which might be said to reverse the sign of the inequality of the usual uncertainty relation.

We shall find that the angular momentum will be only some 2 to 3 % of the momentum of all photons. This worried me because I thought that conservation of momentum of the photon was required. It cannot be because, if it were, the energy balance would be far more than the energy of the photon. Heisenberg resisted the idea that Schrödinger's wave equation was equivalent to his wave mechanics. In the course of time P.A.M. Dirac formulated the matrix mechanics of Heisenberg into his more compact and efficient "bra", "ket" formalism which became a rival to the Schrödinger wave equation. Dirac's formulation of an electrodynamic field theory suffered from the problem of infinities. He also characterized the field in the atom in electrodynamic fashion disregarding gravity. One could express the wave equation using either the Hamiltonian or Lagrangian formalism. It was Dirac who popularly insisted that gravity was insignificant because he looked at the ratio of forces whose fields were based on dissimilar test particles. As we noted earlier had he looked at the fields on the basis of the correct mass test particle he would have realized that using a test particle of a kilogram for gravity was an incongruent comparison. As for the test entity of the electric field it would have been a coulomb. There is no coulomb of charge orbiting hydrogen or any other atom.

Using only conventional masses here is therefore no doubt that GM=1836.15 G is the correct value of the Kepler's law constant for the hydrogen atom.

$$g = \frac{F}{9.1e-31} = \frac{G(1.27e-27)}{r^2\ 9.1e-31} = \frac{G\ 1836.15}{r^2} \qquad 2\text{-}6$$

The Lagrangian is a corruption of the brachistrone, the curve of minimal time descent described by Newton. The Bernoulli's employed it as well. It and the Hamiltonian are used extensively in quantum mechanics. It was used by Schrödinger in formulating his equation. When Friedman derived Einstein's General Relativity in three dimensions he used the Lagrangian at the start which may be a signal that if General Relativity relies on the Lagrangian then it becomes more reliable at an extremely large orbit where changing orbit may be more nearly like free fall. In classical orbit m g r = m v² and it is the kinetic energy ½ m v² that is employed in calculating escape velocities. In other words the sum of potential and kinetic energies is not constant in orbit as it is in free fall. Perhaps entering or changing orbit seems related to free fall in cosmic cases; but, it seems less likely that it would be apropos for a tiny atomic orbit. I did not employ the Lagrangian in deriving the Bessel Laplacian even though I related it to the radial solution to Schrödinger's equation which *was* derived using the Lagrangian. I assume that the effect of the Lagrangian is slight in the radial portion of the Schrödinger equation.

In any event, I can add that d'Alembert and Euler in the 1700's derived a wave equation containing a quadratic from consideration of a Lagrangian. Their wave equation had a quadratic term similar to the Poisson portion; but, did not have the time average portion of my Bessel Laplacian. The d' Alembertian is considered to be a wave equation. That a quadratic could describe a Poisson in a wave equation must not be bar to the applicability of the radial portion of Schrödinger's equation to the atom which I propose. The Poisson occurring as a manifestation of the wave equation is an important finding. The first two terms of my Bessel equation can be expressed as a quadratic given that the remaining terms are zero. Those remaining terms constitute the steady state or time independent portion of Maxwell's equation while the quadratic which is defined by equating the time independent portion to zero constitutes the Poisson. By re-incorporating the third derivative we are recovering a cyclic variable time solution. We will however use the easier $\pm 1/2(N_t - N_t)$ method when we calculate eccentricities. The parametric radial solution to the Bessel equation gives the change in radius upon addition of energy, while the Poisson enhances both the charge in the term in which the electron is found and the central mass of Newtonian gravity. The Poisson acts on the Newton's constant times the mass because they are replaceable by Coulomb's law constant times a constant. Earlier I did not apply the Poisson to the gravimetric portion of the unified field in my book *Quantum Gravity and the Unified Field* in 2000. As is seen in chapter one, there is an equal expressible electric field for every gravitational field. Currently I believe that the Poisson applies to both fields. The book in 2000 was dedicated to my wife Mafalda, then living, and our daughter Pamela whom we lost when she was eight years of age.

The equation that I derived for the solvent extraction and then interpreted as a solution to hydrogen and other atomic radii were confirmed characterized as a Bessel function in cylindrical coordinates by my Mathematics Professor at Indiana State University. Later I could and did derive the Poisson portion from the radial solution to Schrodinger's equation. In atoms beyond helium the radii of *2s* and beyond orbitals added. The technique worked for the addition of "layers" of added orbitals in calculating radii of the atoms of the table; but, I believe that the electron at greater velocity in hydrogen and helium spectra should spiral down.

Niels Bohr established his Institute in Copenhagen and had a parade of visitors. His aid, Kramers, was a student of Ehrenfest. Ehrenfest thus had a good relationship with Bohr and was also a close friend of Einstein. He made minor contributions to theory; but, he was not as great a contributor as he wished and the conflict between his friends Bohr and Einstein and the plight of his son were strains that no doubt contributed to his ending his own life. Ludwig Boltzman, the man who gave statistical mechanics the equation of entropy, also took his own life probably, in part, because he had difficulties in convincing others of his work; although, this type of circumstance happened all too frequently in the time between the two world wars in that area of the world. In 1932 von Neuman showed by a series of axioms that quantum mechanics was *acausal*. Born had devised the probabilistic approach to both Heisenberg's and Schrödinger's work. Einstein was distancing himself from the developing quantum theory that he had helped to instigate.

There arose a thought that there might be a hidden variable theory that would work. Einstein was a proponent of the "Hidden Variable", a term for variables hidden to quantum mechanics as was the unified field and exact atomic radii. There was also a movement even a "proof" that stated that there could be no hidden variables in quantum mechanics. This merely reiterated von Neuman's findings regarding the lack of causality in Quantum Mechanics. It is also a good reason for my picking another name for my theory. Attribute mechanics seemed apropos because the *attribute* connotes a *quality* as opposed to a quantity. Since quantum mechanics very often contends with electrons extra to their atoms it is not surprising that it has been eminently successful in the electronic arena. It has made a huge worldwide technological impact. The matter-wave nature of light and electrons that was proposed by de Broglie had its success in those relations exterior to the atom. However, the de Broglie relation does not apply to the uptake of photons by the atom. The photon and the atom are not exactly a conservative system regarding angular momentum.

Einstein and Bohr at the Solvay conference engaged in heated debate over the completeness of the quantum theory. Apparently at the time Bohr came off best; however, the thinking now tends toward believing that Einstein was correct. Bohr's ideas harkened back to the time of the Greek thinkers who, in the absence of more certain knowledge, *decided* what concept was to govern what. This type of philosophical endeavor, dear to the Greeks, was forgotten in the west until it was re-introduced by the Arabs. From them it came to be used by St. Thomas Aquinas in the guise of Scholasticism. What was known as Scholasticism flourished until Francis Bacon championed the doing of experiments. Experimentation found great practitioners in the persons of Kepler, Galileo, and Newton. Later Bohr developed his two philosophical epistemologies which he described as the Complementarity and the Correspondence Principles. The issue was causality; and, what Bohr termed complementarity meant that the velocity and position coordinates were complementary variables as were energy and time. This eventually became known as the Copenhagen interpretation of quantum mechanics. The notion is really the definition of epistemology itself and can be and was later applied to such diverse scientific phenomena as thermodynamics and biology where the limitation of human knowledge is strained. Bohr claimed by the *Complementarity Principle* that the postulate of the indivisibility of the quantum demanded not only a finite interaction between the object and the measuring instrument, but also created definite latitude in our account of this interaction. Thus waves and particles were tied together. This in effect gave rise to the concept of non-locality which is confusing to say the least. The notion of non-locality persists today and holds hope for construction of supercomputers. The Correspondence Principle anticipates the approach to classicality at large quantum numbers.

Bohr was not enamored by Einstein's photon as if it were a quantum. The nature of the wave particle duality was implied here because one measured what one could. It is easy to see that the non-commutivity of variables, (or the using of operators) is related to the controversy as to whether physical reality is concentrated at a point or spread out over something called phase space. For Bohr, the quantum mechanical indeterminism was a consequence of Einstein's photonic particle which epitomized wave particle duality.

The impossibility of a strict simultaneous spatiotemporal and causal description led to Bohr's renunciation of any classical mode of description. Einstein seriously and righteously objected to this renunciation of causality. Pauli took both position and momentum to be complementary which is intrinsically different from the complementarity between space-time and causality. Position and momentum must necessarily be defined in the physics of a classical state. Complementarity between particle and wave is a complementarity of still another kind which is not deducible from the others. The postulate of non-locality and hidden variables, hidden that is to quantum mechanics, was advanced by de Broglie and Bohm and others over the years.

Bell proposed a theorem in the form of an inequality that supported von Neuman's proof that there were no hidden variables in quantum mechanics. It was later pointed out that von Neuman had assumed what he was trying to prove. Bell's theory has led quantum mechanical thought to embrace the phenomenon of entanglement which currently favorably anticipates the feasibility of quantum super computers. Again these phenomena refer to photons in a wave after they leave the atom or to electron waves attributed to electrons ejected from an ensemble. The term non-locality is undoubtedly occasioned by measurements on an atom of the ensemble of which an emanating wave cannot be located as to its position in the ensemble. Such fragmentation is not the subject of attribute mechanics. Entanglement, a situation developed through quantum mechanics is much currently under study for use in high speed supercomputers.

The Correspondence Principle insists that a more general theory, such as relativity or quantum mechanics can give the same results as a more restricted theory, such as classical mechanics seems to have become. Einstein contended that the theory determined what could be measured. To Bohr his Correspondence Principle meant that *corresponding to a wave there* must be some measurable variable that could correlate the observation albeit with "uncertainty". He, of course, did not care for Einstein's photon; but, he was enthusiastic about the quantum jump to the dismay of Schrödinger. General relativity is conceded to be a classical theory while quantum mechanics is not. Strangely both theories employ the Lagrangian. I assume it is the

29

use of phase space that is the root reason why the two theories are mathematically incompatible. Reconciliation of these two theories is said to be under the banner of a theory of "quantum gravity". A word here is needed to assert the abrasion that quantum mechanics has made of the quantum in terms of the failure of the Schrödinger equation to be exactly solvable for the hydrogen atom. The notion of n loops between two fastened ends of a wave has no defensible relation to the number of variable wavelengths of photons exciting an atom of hydrogen or helium. The harmonic oscillator is therefore amenable to the Lagrangian.

There are excellent accounts of the nuances of thought by both the major and minor players of that early era in books by Max Jammer, Louisa Gilder, and Max Brooks. One gets an intuitive impression of the tortuous mental exercises that the proponents of rival theories went through to secure acceptance of their points of view. I indicate what I have given here to again underscore the idea that the real difficulty in applying classical theory was the inability to solve the radial portion the Schrödinger wave equation for radii. Bohr however had correctly given the radius for hydrogen to be .529 Å. His assertion of radii which were integral multiples of that value did not hold. He also initiated the planetary picture of the atom, which is the easiest way to introduce chemistry to beginning students. In beginning chemistry introduction to the radii of all of the atoms is usually given as it is possible to measure these radii with great precision by means independent of quantum mechanical considerations. It is no great leap of imagination to postulate that the electrons are in layers, albeit of unknown inner expanse, about the central nucleus. Indeed my use of the aufbau principle expressed this in the form of orbits of the real principal quantum numbers, N1s, and integral sub orbitals s, p, d, f, and g. Attribute Mechanics celebrates an *integral* change in the real numbered attributes throughout all sub-orbitals filled or not. There is a layering effect in which the second outer layer radially is not as deep as the first and so on; but, provides the increase in velocity needed to reflect the greater energy to accommodate an outer placement of the electron. To facilitate this I assumed that the positive nuclear charge on the proton, according to Gauss' law, was felt on the surface of the next higher inner orbit. The prior difficulty encountered in estimating the *repulsions of the electrons* can be obviated by first of all being able to calculate the *Poisson density* and second by the assumptions of the natural avoidance of electrons on an equipotential surface.

Bohr's original scheme to depict negative energy levels of photon emissions of the Rydberg series as $-13.6\,ev/n^2$ did not stand up. In hydrogen and helium spectra the added energy reduces radii. Such negative energies were not good constructions. Given the base values of .529 and .315 Å for hydrogen and helium one may employ the parametric equations to find their 1s attribute N. Instead of negative energy the starting point for reduction or radius was available by independent measurement. However, Bohr did derive the Rydberg constant from known variables which was a good but not an exact correlation. Only when the successive change of the photons became unit photons did the Rydberg equation become universally applicable. My iterations are always done by changing the spreadsheet value of an attribute N. And, when the one sided Rydberg expression is first employed it allows exact iteration of the *radial change* of both hydrogen and helium due to the entrance of the energy of the first photon. One changes N until the energy increase of the electron matches the spectral photon energy. The spreadsheet change in radius is the reduction in the base or zero point radii of .529 Å for hydrogen and .315 Å for helium. This illustrates the use of the parametric equations to calculate via the unified field the *change in radii* associated with energy input using a spreadsheet. Once the set of spectral radial changes is found they are used to find a set of reduced the base radii. Now a second iteration of N is done to find the N associated with each member of the set of reduced radii. This second set of iterated N lead to the arguments for calculating eccentricities. We must note now that in helium one photon moves two electrons at a time at least until first ionization occurs.

$$\frac{1}{\lambda} = R\frac{1}{n^2} \qquad \text{2-7-a}$$

I assumed in 1984 that the unified field had first to be additive; second, it had to have the same units; lastly and, most importantly, the mass unit had to be the ratio 1836.15 to 1. Coupled with my ability to describe radii and the postulate of the electron density I obtained good correlation. Sometime before 2000 I had found an independent confirmation of the first and second assumptions. It came from the rearrangement of the Rydberg equation that Bohr had expressed in conventional units which must be expressed in electron units. Galileo discovered that the form of a relationship depended on scale. When you have a central mass in kilograms a unit test kilogram will rotate about its center of gravity. Lower the central mass below the unit kilogram and the one time unit kilogram test mass will become central. Newton's Gravitational constant contains units of reciprocal kilograms. Kepler's law, GM, has the mass ratio of kilograms central to the unit kilogram in the reciprocal unit of Newton's G. It is a matter of scale to retain the ratio of central to test mass in hydrogen to the 1836.15 to 1. P.A.M. Dirac's characterization of the ratio of the electric to gravitational forces at face value was deceiving. Had he compared the fields he would have easily noted that one cannot meaningfully compare the Newton per Coulomb ratio to the Newton per kilogram ratio because there is no entity in the atom with a mass of one kilogram and a charge of one coulomb. As I said, I did send him my 1984 conjecture detailing my assumed unified field when he was at the University of Florida; but, he did not reply. Later I derived the result from equation 2-4, and the Rydberg rearrangement.

It is known that for any elliptical motion there is a circular motion with the same frequency. We also must assume that even though the electrons are in elliptical motion there is a time average radius for the atom which, as it is measured, yields a reproducible and acceptable macroscopic measurement of volume.

Also, it is important to consider the notion of spin, which still remains somewhat of a dilemma because there are two ways an electron can spin. In an elliptical orbit, with the central mass at one focus, the force producing that orbit is what is known as spin. It is considered to be magnetic, a reaction consisting of a polar inversion of one or both the orbiting electrons every half revolution in going from ap-atom to peri-atom. Each electron's axis thus inverts, each of them still spinning on that axis. Each half completion of orbit is accompanied by a change in the Poisson, a function of "N". This is how I conceptualize spin. Each spinning electron has half of an attribute number, N, to maneuver. You cannot think of it as half of a quantum number, n, as well. The derivative of the Poisson does not utilize all this space. Since there is only twice the ±½ an integral "n" between shells a spinning elliptical orbit has to accommodate itself within this space. To salvage the spin up-spin down convention a rotation characterized by such an inversion of up-down radial extensions is, I should think, a most likely possibility. Spin on the electron's axis is a topic of another chapter.

The pictures of the p, d, and f orbitals as usually given are better accommodated in my model as a different style p orbital. These novel p orbitals have a pole each side of the one central pole accommodating six electrons. The usual three mutually perpendicular p orbitals have six crossing points. If the poles are set 45 degrees apart the electrons still have potentially six ways to cross but they are in two places. If one staggers the electrons in orbit, crossing collisions are obviated. The d and f orbitals can have 5 and 7 poles 30 and 22.5 degrees apart. An f orbital has six orbits each side of a central one and can accommodate fourteen electrons. It may stretch it too much to have f orbitals with three poles each side of a central pole; but this is where a g orbital originally for super-heavy elements can provide relief for the last of the actinides. One can devise such a pattern that does not harbor conflicting electrons by staggering the electrons arriving at the two crossing points since there are but two. So with the elliptical path the electrons can move to avoid one another presumably obviating the occurrence of the electrons of one sphere having a collision with electrons of another sphere. It may be necessary later to rethink pro rating orbital radii for occupancy.

Louis de Broglie maintained that there were "two clocks" in the every atom. He spoke of a pilot matter wave. He thus became involved with the hidden variable debacle. This is a good place to remark about the entering photon. Every photon ostensibly has the same angular momentum h, or the units of Planck's constant. If $M_0 c^2 = h f$, then the energy of the photon has a mass as well as a waveform. It can't transfer both energy and mass. So it has no matter to transfer to an electron. A photon is not considered to have mass anyway. Consider angular momentum of the photon, h. There is a loss of angular momentum, a broken symmetry. Even if the mass of the photon were allowed an orbit along the path of the electron traveling only slightly below the speed of light then such an occurrence would disrupt the energy transfer if it were sufficient to conserve the "angular momentum" of the entering photons. The problem of this is that the energy balance goes wildly off high. I mentioned that only 2-3 % of the supposed angular momentum of the photon shows up in the electron orbit. Therefore my derivation of the de Broglie equation from the LHS of the Rydberg Rearrangement actually *returns only the units of Planck's constant as it would apply to solar cases*. When entering the atom the quantum number n has no role other than unity because it is just the counting number of entering photons as mentioned in chapter one. In this vein of thought the de Broglie relation only asserts that one Planck's constant has units of angular momentum. This being said we have little hope for a *solar integral quantum gravity*.

The Noether current now has to be an *assumed* current. Such a current has been credited with being active in the atom; but, it is a feature concomitant with the Lagrangian. The Lagrangian is not supported in Newtonian systems. If the Noether theorem is not supported then quantum mechanics is not supported either. When the photon is released from the atom its angular momentum, if it were a reality, is evidently restored. There is no symmetry between a photon and an electron. There is of course conservation of angular momentum that occurs within atomic orbit even if it follows an elliptical path. Under those conditions where angular momentum of the photon were to be conserved surely there would be some mass that represents the photon mass, an Mo, which would accompany the photon in its escape to the atom, so that we could confirm de Broglie's contention of two clocks in the atom. The energy of this would confound the energy balance. It is *not angular momentum* but rather *energy* that transfers. *All this kills conservation of angular momentum and de Broglie's theorem.* Consider as proof that energy can transfer into the atom by energy, i.e. the $h\,f$ of photons or by electric current. Yet these energies transfer out as photons presumably with momentum h *even when excited by current in a spectrum tube.* We cannot accept n h = m v r, the usual statement of de Broglie's hypothesis within the atom. When the quantum number advances it is really only counting the number of entering photons, a process which is totally unrelated to the angular momentum of the electron. Thus quantum number, n, is not counting entering momenta. Obviously then there is no symmetry between the atom and a photon. I must conclude that I cannot support the de Broglie hypotheses because I calculate too many wavelengths parading evenly around short of closing my radial path. The de Broglie's hypothesis, *nhr=mvr*, can be derived from $n\hbar V = Gq$; but, it only specifies h = mvr for n=1. It can refer to the solar system when the macroscopic value of Planck's constant is applied but n is no longer integral. There is no wave on an elliptical orbit needing to have a descriptor having an integral number of n. One of Planck's dilemmas was in stating a classical concept combined with a non-classical concept. If this is still attributable to the variable h then this is the price to be paid for releasing the energy of a photon to an electron in an atom. Strangely any angular momentum that the photon had is conserved when the energy emerges.

In chapter one I was able to show planetary quantum numbers based on an equation similar to the de Broglie hypothesis using a macroscopic value of Planck's constant. The exact solar non-integral quantum numbers are not assured because there are possibly connections to precession yet to be explored. These quantum numbers were real, not integral, numbers ranging from 1 to 20 for the planets. In my 2000 book, *Quantum Gravity and the Unified Field*, I used the geometric mean potentials of transition metals to calculate the known speed of sound in these metals by employing the WKB approximation to the Schrödinger wave equation often used in cosmology. These potentials will now change somewhat and it remains to be seen if the calculation can still be of interest. There also remains a good possibility that average Poisson potentials may be indicative of the Pauling scale of electronegativity. This has implication for the use of the rare earths in photovoltaic systems and other condensed matter physics as well as for biologicals and medicine.

Hopefully I have appealed to the intuitive nature of physical phenomena just as Newton did when he modeled the solar system. An anomaly already addressed is that the solar system has a test mass of one kilogram which results in a gravitational field expressed as Newtons/kg while the electron mass is the test particle for the atomic case. Surely we do not see planets changing orbit due to photons from outer space. There is no greater energy source in any solar system than its sun. Temperature has little if any effect on planetary gravity. There is indication that temperature will change the value of the *principal attribute*, the number that codes the first orbital of any atom. This must happen because we know that the radius changes with temperature. Such anomalies may be addressed using data for linear expansion for solids and volumetric expansion for liquids.

Before presenting actual radial solutions via the parametric equations and velocities by the unified field I would like to explain how the results will be used. One will find in the next chapter a distribution of radii versus attribute, N, that has the shape of a Gaussian distribution. Since N is the argument of the Poisson it is probably better called a Poisson distribution. Only the right hand side of the maximum is utilized. The distribution ranges from a maximum slightly greater than .529 Å at an *attribute* of 40.5 to zero Å at infinite *attribute*. The *attribute* is a real variable, N, which varies by an integral 5 units between shells and one integral unit between subshells. We will first concentrate on hydrogen and helium whose spectra are both not exactly covered by the conventional Rydberg treatment. When the hydrogen electron or helium electrons jumps, it moves to orbit radially less than the previous. I think that the jump is really a smooth transition which would please Schrödinger. The initial value of N for the hydrogen radius of .529 Å is approximately 40.54. When it comes to helium the initial value of N is 43.5. The two electrons will move as a unit in helium from a starting radius of .315 Å. There are two major breaks in the helium ionization, one presumably at the first and another at the second ionization constant.

Unless you assume that the helium electrons move as a unit there would be great difficulty in assigning movement at all. The starting radius in helium is assumed to be from circular orbit. Even when there are two atoms as in the first, or 1s, orbital the value of the calculated radius need not be doubled; but the spectral energy is applied to both electrons until one departs by ionization. In atoms of series beyond helium the volume of remaining electrons after loss of the valence electrons are indicative of ionic radii. Second ionization spectra can be calculated; although, the line of demarcation will not be distinct. In other words ionization of sodium for example may mean that only the 1s orbital electron moves.

Two further considerations must be noted. For multiple orbital atoms such as aluminum, for example, there is a unique value of N for the 1s orbital which I term the principal attribute. The name *attribute* was chosen to pertain to quality, the work quantity having been used to describe quantum mechanics. Given the principal attribute of an element, which is an iterated real number, the values of subsequent s-orbital quantum numbers, n, increase by five. The numbers 0, 1, 2, 3, 4, are the changes in the principal quantum number, n, reserved for the l quantum numbers indicating the s, p, d, f, orbitals and lastly I added a g orbital as earlier described. Only if you are interested in the second half of the actinides the g orbital might be of interest. The iteration was done so that the sum of the radial values when doubled equals the published radius for the atom in question. This was done for you in a Table of Principle Attributes in the Appendix. The table contains atomic numbers, masses and known radii of the elements. If you are doing serious work you may want to confirm your results; or, if you would want to alter the way you treat unfilled orbitals you have the starting points needed to confirm the principal attribute. It is easy to get an orbital out of place in an unfamiliar area of the table. Using these principal attributes, N, we thereby arrive at a similar matrix of radial values in Angstroms by operating on the matrix elements N using the parametric equations for the radius which has two variables, Alpha and the attribute N. The various other formulas can be inserted to complete the spreadsheet should you wish to review the spectra of hydrogen and helium. In the latter case you will find spectra in the Handbook of Physics and Chemistry.

We calculate a Poisson value for each value of radius by a quadratic formula with one input variable the N value of the attribute matrix $\{ \nabla(\phi) = (9N^2 - 273N + 588)/49 \}$. We then form the unified field expression dividing the unified field numerical expression by the radius *in meters squared* for each matrix entry. The value is now embellished by the Poisson and we can calculate the unified field constant $g_{unified}$ from which electron velocities are easily calculated since the velocity is the square root of the radius times the unified constant. ($v = \sqrt{rg_u}$). It is a greater task, when doing spectra of hydrogen and helium, to iterate and find, from the total energy added by incoming photons, the velocity of the aggregate energy and the radius where the electron resides. Using the radii of the $\pm \frac{1}{2}N$, which must be iterated, the eccentricities may be found. Note that changes in N give only changes in radius. A second iteration of N is required to fine the N values required to calculate eccentricities of hydrogen and helium.

When the *radii found for atoms beyond hydrogen and helium by iteration are plotted versus atomic number* the usual periodicity is found; and if one plots *principal attributes* versus atomic number there results a new periodicity that is *in phase opposite the often published plot of the periodicity of radii*. That is if one plots the 1s N attributes versus atomic number one finds a periodicity that is the mirror image of the usually described radial periodicity.

Quantum Mechanics will always have its own area of expertise. This book only touches the surface of uses for attribute mechanics in photovoltaics, biology, and medicine. Poisson potentials mirror Pauling electronegativity and may be better measure of reaction dynamics in exotic materials. I think that the mathematics of attributes is much easier than that of Quantum Mechanics and may even shorten the time greatly compared to that now required for graduate students who are learning the mathematics they need for Quantum Mechanics. Neil Bohr's interpretation of negative energy did not adequately picture the reality of the Rydberg Spectra and the scheme that worked for hydrogen did not succeed for helium.

Earlier I had found ways to calculate bond lengths using matrix additions of reacting atoms. The methods were not as rigorous as I would have liked and required some selection rules that are rather arbitrary. There was also an extension of the method to do matrix calculation of bond energies. Bond energy calculation was also so arbitrary as to be of only limited use.

It is possible to use one-third "electron or proton entities" to describe quarks and electrons as orbital systems. A neutron becomes three "electrinos" orbited by three "positrinos". A proton is depicted as three "electrinos" orbited by six "positrinos". Upon reflection this means a neutron is still a DUD and a proton is UDU as usual. A downquark is an electrino of charge -1/3 and an upquark is an "electrino" orbited by three "positrinos" or +2/3. With a suitable constant for radius, one finds velocities near the speed of light which yield masses of the particles of the zoo. Using mev of relativistic mass one can iterate most particles of the zoo. This provides entry into the weak and strong force calculations which will further distance the unified field from the Standard Model.

The minimal use of equations compared to quantum mechanical books should attract a wide range of readers. The 2000 book had the "basics", it also had many typos and inaccuracies but, hopefully most will be resolved in this book. When I read the book Einstein's Mistakes: The Human Failings of Genius, a Google book by Hans C. Ohanian detailing the mistakes of Einstein I was humbled. I recall my physical science teacher Dr. Herschel Hunt at Purdue saying that he thought that no one could ever be too humble. If you know physical chemistry teachers perhaps you can understand his point. The importance of this work here is that the classical view has been reinstated. Causality is vindicated which would please Einstein greatly. There has been widespread belief that at some dimension that gravity and electromagnetic theory would become the same. This for some time has been believed to occur only at the Planck Length scale, an unbelievably small dimension far from our everyday use. Now I can say that the line of demarcation, if any, is between the solar and atomic cases. In chapter one I indicated that we are not yet done with the Rydberg Rearrangement as we shall see in chapter three. It is said that you haven't lived unless you have been wrong at the top of your voice.

Recall from chapter one that we know planetary radii and gravitational acceleration with respect to the sun. Take the protons per kilogram times the mass of the sun and you have the positive charge on the sun. Using Coulomb's law, find the negative planetary charge resulting from this positive charge of the sun which will yield the known planetary acceleration. Surprisingly, when we divide that negative charge per planet by the mass of each planet we get the same value for every planet. Conclusion: there is an analogue of Newtonian gravity that can be expressed by Coulomb's law. This is an entry point into cosmology. This is the fact that allows us to place Coulomb's law constant into general relativity. In fact the concept of dark energy stems from the too rapid orbital velocities of certain central masses which appear to have insufficient mass to sustain. If the earth were sprayed with electrons emanating from a neutron star our negative charge to mass ratio would6 go up and we would be more strongly attracted to our star the sun with a period of orbit less than a year. There has been speculation that G in other parts of the universe is not the same as we have in our solar system. Now we have a mechanism that celebrates this.

Several projects are being pursued by persons of considerable educational expertise accompanied by, I am told, billions of dollars in grant money. This is the case not just in the United States, but also throughout the world; and, with the financial support of all the major world powers. Perhaps the least funded is the attempt to reconcile Quantum Mechanics with General Relativity. This is known as the "problem of Quantum Gravity". The leading candidate is a mechanism called "string theory" which is acknowledged by the world's leading mathematician, Roger Penrose, at Oxford to have serious difficulties. He also does not place much credence in any other current theories. He does say that, of the two theories, quantum mechanics suffer the most change in the reconciliation. I surely can confirm this. A second search was for the hypothetical "Higgs Boson". This extant entity has been found at CERN. Its role is rather nebulous. Its credence depends on predictions of the Standard Model. With the roll of the Standard Model in jeopardy because of its quantum mechanical basis I think that my method deserves a revue.

A third search is the search for "dark matter"; and, finally there is the mysterious "dark energy" which accelerates our outer universe ever faster the further out it is. But wait. Aren't we as far out in an infinite universe as anyone else? Taking these projects in reverse order it is believed that the density of the universe is not easy to figure; but, if known, this density would tell us if we are to stay the same, expand, or experience much later, a big crunch as the matter gravitates and returns to wherever it started from at the big bang. I have reported a density calculation of the universe calculation from Einstein's $E = m\,c^2$ involving Cantor's hierarchy of infinites that shows a slight slow tendency toward the big crunch and is in ballpark agreement with another estimate that involves an assumption of one hydrogen atom per square meter. Obviously this estimate could vary widely from the basic assumption of my estimate which was that hydrogen was would form a cube of one meter throughout the universe. The density of hydrogen could be less but it probably is not greater than this.

The constant that Hubble arrived at convinced Einstein that something was needed to accommodate his apparent outgoing velocities. This occasioned what Einstein termed his greatest blunder, his insertion of the cosmological constant into general relativity. This is most evident in the Friedman equation. Today most argue that the cosmological constant is needed. Despite this the galaxies in the outer fringes of space seem to be traveling outward. I can only conjecture that this outward movement is not universal and that it is more prevalent in certain areas of the sky. It could be that the central masses of these galaxies are greatly permeated with electrons which cause the repulsion attributed to dark energy. I suspect those sending missions to Mars are aware of such difficulties.

I have already commented profusely on the "dark matter" project. One solution of my rearrangement of the Bohr-Rydberg equation predicts that gravity has an electromagnetic analogue. The possession of a greater negative charge on one of two spheres will affect their gravitational attraction. That gravity has an electromagnetic origin has long been suspected and it introduces Maxwell's equations into the gravitational mixture in a natural way. Certainly if one of two metal balls receives a negative charge it should "gravitate" slightly more to the other than normal. We know if the negative charge is great and equally distributed that there is repulsion. Paper bits attached to a van der Graf scatter. Because certain neutron stars spew forth electrons it is highly likely that there would be an imbalance of negative electric charge that will produce the faster than expected orbiting if found in certain areas of the universe. Since Newtonian gravity is sufficiently accurate for our solar (star) system. It would seem that General relativity would be required only to perceive from our system what is happening elsewhere; yet elsewhere should be adequately described by Newtonian gravity if only we were there. But the value of G will be altered if and wherever an imbalance of electrons occurs. We must greatly note that even if G remains the same, "local time" depends on central mass which is indicative of the problem.

Finding of the hypothetical Higgs Boson, presumably the only one of four bosons with mass, was the attempt to make complete the standard model and complete the sense of the eightfold way. Noether's theorem concerns symmetry and conservation. Much has been learned about symmetry and there are many uses both actual and presumed. This Higgs Boson was needed to round out the symmetry of the eight-fold way particles of the zoo. The rationale for the Higgs boson, however, depends on the Schrödinger formalism. Hopefully, the Higgs can have a role in the attribute formalism as well. When Noether's theorem is nested on differentiable (ostensibly Lagrangian functions) the need for massive bosons to conserve symmetries other than energies is questionable. Undoubtedly the role of charge and its incorporation into mass is still of great interest.

The strong force holds the quarks together. I can define relativistic patterns of orbit by increasing the attribute during which the *relativistic mass* expressed in mev, attributed to these entities is reached by high N. This is the way the masses of tiny particles such as quarks, mesons and electrons are expressed. I can increases the velocity in orbit of both the positrinos of protons, or electrinos of electrons, as well as orbiting entities of other zoo sub particles. The more energy one puts into the orbit the nearer the speed of light the orbiting charges go and the greater their observable (mev) mass.

Left to consider is the reconciliation of Quantum Mechanics and General Relativity. Most investigators favor finding a quantum theory of General Relativity. String theory has too many dimensions. There should be an end to the idea of reconciling quantum mechanics to general relativity. Time and money are being spent in finding cosmological quanta. We now have two classical theories. I would like to believe that general relativity and attribute mechanics are compatible if only because of the dual intent of the Rydberg Rearrangement.

The Russian physicist Friedman derived Einstein's General relativity with recourse only to Newtonian theory. He expressed his result in conventional spherical form. When Einstein placed his geometry in this form by reduction of his 4-D space time geometry to 3-D geometry he arrived at essentially the same formula. If you observe the first term you will find the $8 \pi \rho$ term which stems from the use of the Lagrangian in the formulation by Friedman. The left hand side is purely the Hubble constant and the term following the $8 \pi \rho$ on the other side is merely half of it. The entire equation devolves into the definition of the Hubble equation which is usually expressed without the Lagrangian. What does this mean? If you have need of spacetime geometry to find useful areas of the cosmos to explore, then use Einstein's General Relativity. We are told that its accuracy is not needed in the solar system. When planning rendezvous with planets, however, it is probably indispensable in the space program.

There can be no doubt about the Special Theory of Relativity as its efficacy is well documented. I think that if Einstein had had in 1950 the hand held calculators that we have today that he would have found the link between gravity and electromagnetic theory that he so strongly felt existed. He too suffered from not knowing the nature of the radial positions of the atoms. Perhaps he really knew what the link was. In 1933 Einstein, in a speech at Oxford, explained that a field theory could accommodate an atomic theory in the true sense of the word without localization of the particles. Such a theory he said would be compatible with general relativity. We are closer to the TOE!

Measurement involving localization of the electron wave took place external to the atom. An electron knocked out of an atom creates entities antithetic to a normal atom. Certainly in the ensemble we cannot localize the particles nor even in a single atom can the electrons be localized; but a time average field is realizable by the methods I describe.

There is another area which retarded the development of atomic theory besides the lack of evaluating radial position. There is scant mention of the phenomenon of the ensemble. In the rush to characterize the electron by the wave equation much effort was on the electron as if it were apart from the atom. Detailing the wave picture of atoms was necessarily done mostly with regard to spectra. Schrödinger contacted Weyl to solve the radial portion of his wave equation. Because their solution involved two quantum numbers they disregarded it.

The rules I will follow in detailing the attribute numbers in the atom differs from the list of the conventional principal quantum number set. The set of old principle quantum numbers going from 1 to 7 have seven attributes each one five greater than the last. Each of the fives add successively a one until the next. This emulates the 0, 1, 2, 3 or *l* values of the s, p, d, and f orbitals. We then have a *matrix of real attributes spaced on the number line because the* 1s orbit is not an integer.

The summation formula, equation 2-7 b, for a matrix of simple parametric equations for radii of a given atom is:

$$r\,\mathring{A} = 2\sum_{N_P}^{N_i}\sum_{N_{P+j}}^{N_{i+j}} \frac{\alpha \ln \alpha}{\sqrt{N_{i,j}+6}}, \text{where } \alpha = \frac{N_{i,j}-35}{N_{i,j}-28}. \qquad \text{2-7 b}$$

The above formula will give the radius of an atom given the correct principal attribute N for the *matrix of attributes in angstroms*. When the radius is used in calculation of the unified field it must be expressed in meters.

There has always been some confusion over the correct definition of spin. It seems that there is agreement that spin involves magnetism. I believe that the electron has (first) to spin on its orbit. There is a second spin on its polar axis. A third spin occurs as its poles reverse every half of the first spin. All are probably magnetic but the third type spin reversing its polar magnetism at each half of its (first) spinning revolution as it goes from peri-atom to ap-atom is the one defining elliptical motion. Any disparity of the orbital focal axis due to the rotation about the center of gravity is not addressed because its location is not important.

Gravity unlike other fields gives shape to the area of the macro-universe in many disparate locations. We have seen that our solar system has a set of real quantum numbers related to the solar macroscopic Planck's constant. There is no reason why other gravitational arenas of the universe should not follow suit. Then there is the grand hierarchy for no doubt which Einstein had in mind with his formulation of general relativity. Certainly no human mind could come up with the locations required to fulfill any theory needing to know the location of all the masses. Bohr would argue for a plethora of wave functions; but, could only do battle with one at a time while Einstein might be satisfied with no wave function. Quantum mechanics is non-classical; general relativity is classical. Quantum attributes affirms the classical nature of general relativity and offers a classical alternative to quantum mechanics for some but not all situations. Slit and beam experiments *are not* the domain of attribute mechanics.

I have Leon Lederman, the former director of Fermilab, on tape saying that our present system for things atomic is "just too complicated". John Archibald Wheeler on the same tape can be observed saying that "when we see the correct solution we will all say that it is so beautiful that how could it have been otherwise". I cannot say how this seems to you but to me, after over fifty some years of musing, the story seems more beautiful than I could have ever hoped for.

Chapter Three

Classical Solution to the Schrödinger Radial Equation

We continue with development of the atomic calculation implications inherent in the equation from chapter one which resulted when the rearrangement of the Rydberg equation was equated to Newton's universal gravitational constant. When we assert a classical solution to Schrödinger's radial equation we mean that his traditional *radial solution*, the Associated Legendre equation which was never previously solved exactly, is *now solved* by virtue of my being able to show an equivalency to my Poisson distribution function. The Poisson is a part of my Bessel function which originally yielded the radial solution. The solution entails a succession of attributes, the new way to represent a quantum number. Three serendipities and two vital observations, starting in the 1960's, combined to allow this chapter to be written. In the order of their occurrence first came the Bessel function, then the radial solution. Finally the guess of the unified field followed by its derivation from the Rydberg Rearrangement. One vital observation was the conclusion that the Poisson was inherent in the Bessel equation. Another vital observation was that one must employ *the unit electron mass* as the test particle, not the kilogram, in the atomic realm. The unified field was never successfully asserted from 1920 until my representation. The radial solution can be expanded in an aufbau matrix form to calculate the published radius of any atom. Since there is no intervention of the velocity in the published radii we avoid complementarity, the correspondence principle and uncertainty.

Einstein tried often in his later years to reconcile gravity with Maxwell's equations but was not successful. He did not suspect that there was a dual solution, one for solar and another for the atom. The often used Lagrangian reverts to a Newtonian system when a *radial distance* is involved as it is in the atom. The use of electromagnetic theory potential fails even in the Schrödinger equation partly because the demand for the unit coulomb as the test particle of the field was so ingrained that the error of such a usage went unnoticed.

The notion of inclusion of an additive gravitational field in the atom which rivals the electric field in hydrogen really points to the fact that there is also an underlying field in the proton and electron as well. Therefore these particles have three negative or three positive q/3 centers with six positive or six negative q/3 orbiting entities respectively. Because we can now write equations for these orbiters in terms of a unified field this casts a shadow of doubt on the need for the standard model.

For now consider the field defined by equation 3-1 b from the RHS of the Rydberg rearrangement of the first chapter. One can consider the radical to be negative unity and multiply both sides of the equation by a Poisson which comes from a quadratic in the value of the attribute, a non-integral number, N, associated with the radius equation 3-1 a. The radial value calculated from the parametric equations (3-1 a) is the value of r to be used in equation 3-1 b.

$$\alpha = \frac{(N-35)}{(N-28)} \qquad r = \frac{\alpha Ln(\alpha)}{\sqrt{N+6}} \qquad \text{3-1 a}$$

$$\frac{G\,M}{r^2} = \pm\frac{KQq}{mr^2}\sqrt{\frac{mvr}{\hbar}} \qquad \text{3-1 b}$$

The $\pm$ sign resulted from taking the square root of the radical which is unity. I exploited the idea of taking the positive sign of the radical which led to the equality of the gravitational *and* electromagnetic fields that was explained in chapter one. We are going to develop the unified field that results from taking the negative of the RHS of equation 3-1 b. This results in the addition of the gravitational and electromagnetic fields which can have no other conclusion than that we no longer have equality but additivity, a novel form, a unified field that also stems from the Rydberg Rearrangement. That each sign of the radical has its own domain points to additional inner centers, also electromagnetic, that gives additional insight into the usually perceived relation between the proton and electron Because the sub atomic gravitational centers undoubtedly has an electromagnetic equivalent I reasoned that the Poisson was applicable to both entities of the unified field. This was the case in the solar system except that the Poisson was necessarily unity. The Bessel equation was observed to have dual forms, a Laplacian when in vector potential form and the Poisson when in scalar potential form. The vector form transforms to the scalar using the definition inherent in the time independent portion. The Bessel equation doubles as a d' Alembertian.

Therefore in this chapter we will consider ramifications of the unified field. The presence of the minus sign occasioned by the ± sign of a radical is reminiscent of what Dirac exploited in order to find antimatter. Here we can attach significance to the minus sign because it allows addition of the electric and gravimetric vectors to form the postulated but heretofore never expressed *unified field of the gravitational and electromagnetic fields*. If the negative connotation of the right hand side is moved to the left of equation 3-1 b where it is positive and adds to the gravitational then we cannot have zero. We can only attach significance to the addition of the two fields as their unification to a meaningful number that has never yet been exploited in the literature. There is left the only conclusion, namely that we have formed the unification of these two field from the rearrangement of the Rydberg equation, an equation that had been eminently successful in correlating hydrogen spectra but not helium spectra. Will it now work for helium spectra? I found that now it does work for both atoms of hydrogen and helium associating radial changes of the electron now with published spectra by accepting the change of energy of spectra and taking on new radii, velocities, Poisson values, and eccentricities associated with the spectral changes.

The reason that our innovation works has to do with the fact that we *omitted all but* $1/n_1^2$ *of the last bracket in formulating the Rydberg Rearrangement*. While n only became equal to unity in the radical *of the Rydberg Rearrangement*, it nevertheless remained as a counter for the entering photons *outside*. This has some repercussions for the word quantum as used in a theory. For one there is nothing to prohibit the uptake of any spectral value in the equations themselves. Can we toss the verb to quantize? It will become doubtful if there is even a starting point or whether a photon from helium could excite a hydrogen atom. Recall that the older restriction for succession was the necessity to fit the observed orders of spectral frequencies of the photon. This now discredited theory also drops conservation of momentum. Noether, after a prolonged study of General Relativity, assumed the Lagrangian in the development of her first equation requiring conservation of angular momuntum.[9] Implication of the conservation is dropped upon reversion to Newton's law when the distance is radial. If the photon has no mass it can have no momentum of any kind. So there is no broken symmetry. This has similar implications for the standard model which also employs the Lagrangian and concerns itself with conservation of angular momentum in phase space. I will show that classical subatomic Newtonian systems are relevant for sub atomic systems as well.

From my data one cannot expect in any way to have angular conservation of momentum because the velocities entailed far exceed the energy balance. If you consider conservation of energy based on the value of energy such that $m_e v^2 = hf$ then there will be associated angular momentum within the new orbit. Using a spreadsheet one needs only to iterate the change in radius concomitant with the new associated field conforming to the velocity change associated with the energy of the photon entering. Knowing the radius of the original atom, .529 Å for hydrogen and .305 Å for helium, the values of radius must be subtracted from the basal radius as each photon energized the electron. In the case of helium one must assume that the two electrons reduce in radius in consort. When the first energy of ionization is reached in helium then the remaining electron receives all the energy of the entering photon. Astonishingly, I found that the radii of the electrons were linear functions of the cumulative velocity which by passes the equation; but, the constants of the linear relations are not known so it is only a curiosity. Since the cumulative change in velocity is a linear function of spectra one may find final radii by their subtraction from the basal radius apparently partially circumventing my process. Unfortunately, the required slopes and intercepts are not known. There is but one equation for hydrogen; but, there are three separate linear relations for helium, two for the first ionization both with the same slope and differing only slightly in intercept, and another for the second ionization.

The difference of two inverted square quantum numbers found in the Rydberg equation, which did not work for helium and was neither entirely accurate nor complete for hydrogen, is not a part of the new correlation. The notion of a jump persists if you want to perceive it as such; but, each jump is correlated *only* by the absorption of energy associated with some nth photon. Actually one must decide whether to admit the lowest or highest energies to enter first. I let the low energies enter first. This may be counter to the Rydberg scenario which seems to accord the Lyman series the first entrants.

If one inverts the order of entering photons and allows the high energies to enter first then the radial values invert as well. On wonders if the atom would not accept any energy proffered. In a spectrum tube the atom gets energy from an electric circuit, instead of a photon. Hopefully it has some selective means of taking energy in the sequence we see in the photons that emanate. Extremely high temperatures can be found with helium spectra approaching 5000 degrees. We will investigate the effect of temperature on the iterating variable in a later chapter. In a spectrum tube that extra energy is easily supplied. In an ensemble my equations do not *demand* an order; however, it seems likely that there could be a preference for order considering the original Rydberg development. Given that some atoms all could be accepting a single energy, and with all other atoms each having like preference for one of the other energies, then these could, as an ensemble, present the same series of spectral lines normally expected. This is, hopefully, not a likely scenario. The other possibility that any atom could pick up any energy presented and with the ensemble radiating the usual spectrum seems more likely. Regarding the any energy pickup scenario, no atom could divulge its overall status because observation could not reveal its previous history.

The Poisson is evaluated by the N, or attribute variable, associated with the radial value r.

$$V^2 = rg_u = \frac{(GM + KQ)}{r} \nabla^2(\phi) \qquad \text{3-2 a}$$

$$\nabla^2(\phi) = (9N^2 - 273N + 588)/49 \qquad \text{3-2 b}$$

---

[9] Dwight Neuenschwander, "Noether's Theorem in the Undergraduate Curriculum", *Am J Phys*, 82, no. 3, (March 2014): 183.

This is accomplished by increasing the attribute, N, until the velocity of equation 3-2 a equals the velocity of the electron indicative of the energy of the product of Planck's constant and the frequency of the entering photon. After exhausting the entering photons one must now take the differing changes in radius in turn *from* the radius of the original atom which yields the successively decreasing values of the velocities of the electrons. This zero point radius is .529 Å for hydrogen and an average .305 Å for two helium electrons. One now has the actual atomic radii and from these *a second set of attributes, N, can be iterated*; but, these have value in that the changes in Poisson determine the eccentricity. One of the most recent and intriguing events occurring was the discovery that the natural logarithms of the entering wavelengths in meters and the first radii entered were linear for both hydrogen and helium and have both slopes and intercepts within 3-sigma limits. As indicated earlier, the slope and intercept would have been unknown. When the radii found from linear equations are subtracted from the zero point values the consequent radii are within .006 % off the values found regularly. Helium, with two electrons, required a doubled use of entering energy until first ionization was reached and then was continued with a lone electron using energy until the spectra were all entered. In addition the linear correlation for helium during the first ionization was split in that the first portion had a very slightly different intercept although the slopes were the same. This is shown in Figures 3-4, 3-8, and 3-10.

Another facet was interesting in that one could start with entering either the high or low energy end; but, if you reversed the order of entry the iterated attributes reversed also. Original time for iteration of helium was about eight hours for some 125 entries; although with practice it could be cut to half that time. I probably did the iteration four or five time in developing the technique. I decided to input the low energy spectra first as that seemed more logical.

The output of the first iteration was a set of differential radii. These radii are used to get succession of reduction of the original zero point value of radius. This is indicative of the downward spiral of the electron as photons enter the atom. Surprisingly the entire calculation could be shortened by using the linear relation between spectral wavelength and differential radii. Constants of the line would not have been known, however. I would not recommend doing this as the reasoning behind the process is circumvented; but the fact is worth presenting because it increases the authenticity of the process. Even the crowning touch to disproving uncertainty, the calculation of eccentricities, was shortened.

To calculate eccentricities two nearby disparities in radii are needed. You obtain these by allowing a half attribute above and below a middle point. This variation is on the iterated attributes of the *second* set of radii. This half attribute variation is surely the most that can be allowed. This is the swing from which to calculate radii that determine maximum eccentricity. Given an ensemble and the percentage of energy emanating at each level one is able to calculate a series of eccentricities.

Helium in particular has to reach an extremely high temperature in order to fully double ionize, perhaps as great as 5500 degrees. The *velocities of atoms* at that temperature have calculable values. We must assume the velocities of the electrons in the atom are independent of the motion of the atom as a whole. The constant amount of energy above that to reach temperature have to fuel the ionization; but, this is not achieved at the expense of the energy entering photons as is evident from the presence of currents fueling the spectrum tubes. I am able to calculate conditions that conventional quantum mechanics has never attempted. Due to the nature of an ensemble one could find the percentage occupation in each energy level by calculating the strengths of the spectral energies in each category without disturbing the system. The result would undoubtedly depend on the current. How quantum mechanics ever presumed to make one measurement on one single unspecified electron in an ensemble and, *in light of uncertainty, figure a probability of anything* is inexplicable.

The value of q, the negative charge on an electron, which would have formed the test particle of an electromagnetic field, is seemingly lost in forming the unified field. However, recall that for every negative charge q, there is one unit electron mass, m. Nevertheless, as we explained in chapter two *the coulomb unit does cancel* and the unit mass conveniently takes its place, WLOG. This is so because the mass products Mm of central and test mass were required to be in electron masses in order to equate the ensemble to G. In other words the unit coulomb per unit electron mass is physical unity. Fortunately I had a radial solution even before I came upon the equation of the *Unified Field*. As a matter of fact I had deduced the field without the Rydberg rearrangement almost a year before. This was fortuitous because I had not been able to use dimensional analysis from conventional solar frameworks of Newton's law to verify my supposition of the unified field. Thus the verification of my jump to using electron mass units of unity via the Rydberg Rearrangement was highly satisfactory because I could now select an appropriate mass unit to render the rearrangement compatible with Newton's law. Amazingly without embellishment by the Poisson the solution would have shown the gravitational field to be some *85 times stronger* than the electric field in room temperature hydrogen. The disparity continued because I realized that I must also allow the Poisson to act on the gravitational field as well. I decided that the Poisson applied to both sides of the equation because the gravitational field has an electromagnetic equivalent. Surely pronouncements regarding the lowly strength of the gravitational field kept analysts from discovering the true nature of the unified field earlier.

The equation that I derived in the sixties is shown below:

$$(N+5)F + 2F' - (N-28)F^{-1} + 7 = 0 \qquad\qquad \text{3-3}$$

It is a Bessel equation in cylindrical coordinates. The first two terms are the Poisson. The remaining terms are orthogonal to the first two and equal zero. They constitute the time independent solution. The solution to atomic radii is given by the parametric equations 3-4 a b:

$$\alpha = \frac{(N-35)}{(N-28)}$$

3-4 a

$$r = \frac{\alpha \, Ln(\alpha)}{\sqrt{N+6}}$$

3-4 b

$$F = \frac{1}{1-\alpha}$$

3-4 c

$$F' = \frac{1}{(1-\alpha)^2}$$

3-4 d

$$\alpha = e^x$$

3-4 e

$$\frac{\alpha}{\rho} = r$$

3-4 f

The first two equations, above, of 3-4 are the parametric equations. They permit plotting the radius versus N which is a Poisson distribution. This is shown in Figure 3-1. It appears to be a Gaussian skewed right. Applying the value of N of equation 3-4 a to equation 3-2 b one may calculate the Poisson. The first two terms of equation 3-3 contain the vector functions F and the derivative of F' of the define the scalar density which is the Poisson when converted to the quadratic pf equation by equation 3-4 a which is obtained from the last two terms of 3-2 b when equated to zero. Their vector nature is demonstrable because the first two terms of 3-3 are orthogonal to the rest of the equation as shown in Figure 3-8. I derived the Bessel function (equation 3-3) in the early 1960's. It required 17 pages of what may be rightly termed Newtonian fluxions; but, it assumes its abbreviated form because the definitions of F and F' are Taylor series in the exponential of equation 3-4 e. Without it we would not have the Poisson as a function of known radius. It is of an added value to derive those first two terms of the equation 3-3 from another source, the radial separation portion of Schrödinger's equation known as the Associated Legendre equation. Not only does it bolster the derivation of the Bessel equation, but it involves Schrödinger's work. Although I had found the radial parametric equations earlier, the following derivation of the Poisson from the Associated Legendre equation lends further support to the nature of a new variable ρ, Schrödinger's radial variable, to my time like variable N of equation 3-4 e.

To derive the first two terms of the Bessel equation we start with the radial portion of Schrödinger's wave equation known as the Associated Legendre equation.[1] Letting B = l (l+1):

$$\left(1-\varepsilon^2\right)\frac{d}{d\varepsilon}\left(\left(1-\varepsilon^2\right)\frac{dP}{d\varepsilon}\right) + B\left(1-\varepsilon^2\right)P = m^2P \qquad \text{3-5}$$

We shall re-derive the Poisson, which we already have from the Bessel equation, this time from the Associated Legendre equation.by first setting the magnetic quantum number, m, equal to zero in equation 3-5. Then we factor out $(1-\varepsilon^2)$ and substitute F for P. We have set up in equation 3-6 a so as to differentiate by $d\varepsilon^2$ instead of $d\varepsilon$. We will set P'' to zero because it is a vector in the time average direction.

$$BF+\frac{2d}{d\varepsilon^2}\left((1-\varepsilon^2)\frac{dF}{d\varepsilon^2}\right)=0 \qquad \text{3-6 a}$$

$$BF+\left((1-\varepsilon^2)\frac{2d}{d\varepsilon^2}\frac{dF}{d\varepsilon^2}+\frac{dF}{d\varepsilon^2}\frac{2d}{d\varepsilon^2}(1-\varepsilon^2)\right)=0 \qquad \text{3-6 b}$$

$$BF+\left(2(1-\varepsilon^2)F\,''+F\,'\frac{2d}{d\varepsilon^2}(1-\varepsilon^2)\right)=0 \qquad \text{3-6 c}$$

$$BF+\left(F\,'\frac{2d}{d\varepsilon^2}(-\varepsilon^2)\right)=0 \qquad \text{3-6 d}$$

$$|BF|+|-2F\,'|=0 \qquad \text{3-6 e}$$

I have taken the absolute values of the two terms in equation 3-6 e because each is a vector, they are orthogonal and their scalar magnitude is not tied to the direction of the derivative of F. When we compare with equation 3-3 note that if we give B the value of N+5 the first two terms of equation 3-3 are identical with 3-6 e. This constitutes the proof that the Poisson portion of the Bessel equation is derivable from the Associated Legendre equation which is also the radial portion of Schrödinger's equation.

The parametric equations are next to be derived. I will now determine the nature of the coefficients of the exponential that determines the value of α. The first two terms of the Bessel equation can be characterized by equation 3-6 f as follows:

$$J\left(k,\rho\right)=(k^2-1)\,F+2F\,'=0\;,\;\;F=\frac{1}{1-e^{f(\rho)}} \qquad \text{3-6 f}$$

Then from equation 3-3 we let $k^2+1=N+5$. $k=\sqrt{N+6}$. Equation 3-6 f has a set of conceivable solutions for the exponential, namely $\{\,k\rho,1/k\rho,\rho/k,k/\rho\,\}$. Alpha must be less than one. At an N of 40 the value of α from (N-35)/N-28) is .42 and the Ln (.42) =-.88. In going from N of 40 to N of 100 the absolute value of the logarithm decreases almost nine fold. Thus for a nine fold decrease in k there must be a six fold decrease in r. This eliminates all but x = k/ρ. Since $k=\sqrt{N+6}$ we get for the exponential for α:

$$\alpha = e^{\frac{\sqrt{N+6}}{\rho}} \; ; \; Ln(\alpha) = \frac{\sqrt{N+6}}{\rho} = \frac{k}{\rho} \qquad \text{3-6 g, h}$$

Why do we need an exponential for alpha? Because the functions F and F' are a Taylor series and its derivative with alpha varying between zero and one. The variable rho is not the radius. One must invert rho in alpha to obtain the radius and so the parametric equations can be found.

$$\rho = \frac{\sqrt{N+6}}{Ln(\alpha)} \; ; \; r = \frac{\alpha}{\rho} \qquad \text{3-6 i, j}$$

Solving for ρ in equation 3-6 j and substituting I into equation 3-6 i, we have the second parametric equation 3-6 l.

$$\alpha = \frac{(N-35)}{(N-28)} \; ; \; r = \frac{\alpha Ln(\alpha)}{\sqrt{N+6}} \qquad \text{3-6 k, l}$$

I assumed that the second derivative was zero.

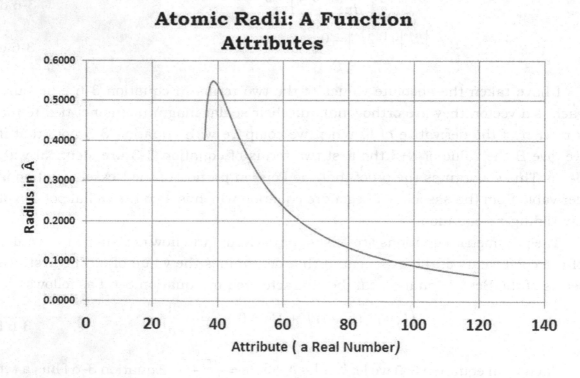

Atomic Radii: A Function Attributes

To obtain the Poisson equation from the parametric equations 3-4 a, b we consider both the first two and last two terms equal to zero. We can now use $F^{-1} = (1-\alpha)$ to solve the last two terms for in terms of N, equation 3-4 a. That Substituted into the first two terms of the Bessel equation gives the Poisson as a quadratic in N. This allows a unique value of the Poisson and radius for *every* value of N. It is known that the vector and scalar fields have the same value in cylindrical coordinates. In the 1700's d' Alembert and Euler derived a wave equation from consideration of a Lagrangian that had the form of a quadratic. This is another assurance that equation 3.3 is a wave equation. The Poisson of the change of radius I found to have the form of equation 3-7:

$$\nabla^2\phi = \frac{(9N^2 - 273N + 588)}{49} \qquad 3\text{-}7$$

The Poisson was given in the first chapter without derivation. It acts in outer space via novae to negatively embellish either Q or q of the Solar Field expressed as an electric field. This is a multiplier of Newton's constant, G, as the two fields are equivalent. The so called questions of dark matter can be resolved by attributing a Poisson to embellish Newton's law in outer space to account for the "lost matter" by excessive electrons most probably residing on orbiting mass. The Poisson has to be unity for the solar system; but, it surely is greater in the "dark matter" areas of outer space.

We may now conclude that our atomic *Unified Field* must take the form:

$$g_u = \frac{GM\nabla^2\phi}{r^2} + \frac{KQ\nabla^2\phi}{r^2 m} \qquad 3\text{-}8$$

$$v^2 = rg = \frac{(GM + KQ)}{r}\nabla^2(\varphi) \qquad 3\text{-}9$$

From equation 3-9 it is apparent that the Poisson affects both elements of the unified field. Using only the parametric equations 3-4 we may plot the attributes, N, versus the 1s radii. It is a Poisson type distribution. In Figure 3-2 the value .529 Å occurs slightly to the right of the maximum at N = 40.54. This is the principal attribute or 1s attribute of hydrogen. A table for the values of *principal 1s attributes* for all atoms of the Periodic Table is given below.

Table of Atomic Attributes* Their Numbers, Radii, & Molecular Weights

		N 1s	R	Mw	©2002			N 1s	R	Mw
1	H	40.54	0.529	1.0079		53	I	112.76	1.33	126.9045
2	He	43.15	0.92	4.0026		84	Xe	114.79	1.31	131.29
3	Li	44.87	1.23	6.941		55	Cs	81.09	2.35	132.9054
4	Be	59.60	0.89	9.0122		56	Ba	92.34	1.98	137.327
5	B	64.62	0.82	10.811		57	La	102.65	1.69	138.9055
6	C	69.15	0.77	12.011		58	Ce	105.07	1.65	140.115
7	N	73.13	0.74	14.0067		59	Pr	105.39	1.64	148.9076
8	O	77.95	0.70	45.9994		60	Nd	105.75	1.64	144.24
9	F	81.79	0.68	18.9984		61	Pm	106.54	1.63	145
10	Ne	85.03	0.67	20.1797		62	Sm	107.33	1.62	150.36
11	Na	57.10	1.54	22.9898		63	Eu	98.88	1.85	151.965

12	Mg	65.95	1.36	24.305	64	Gd	108.06	1.62	157.25
13	Al	72.77	1.18	26.9815	65	Tb	108.87	1.61	158.9523
14	Si	76.83	1.11	28.0855	66	Dy	109.68	1.6	162.5
15	P	90.42	1.06	30.9738	67	Ho	110.93	1.58	164.9303
16	S	83.82	1.02	32.066	68	Er	111.32	1.56	167.26
17	Cl	86.90	0.99	35.4527	69	Tm	111.68	1.58	168.9342
18	Ar	89.02	0.98	39.948	70	Yb	106.88	1.7	173.64
19	K	61.30	2.03	39.0983	71	Lu	113.37	1.56	174.967
20	Ca	73.35	1.74	40.078	72	Hf	119.94	1.44	178.49
21	Sc	78.47	1.44	44.9559	73	Ta	128.25	1.34	180.9479
22	Ti	93.33	1.32	47.88	74	W	129.34	1.3	183.85
23	V	88.14	1.22	50.9415	75	Re	131.24	1.25	186.207
24	Cr	90.72	1.18	51.9961	76	Os	126.70	1.56	190.2
25	Mn	91.99	1.17	54.938	77	Ir	133.07	1.27	192.22
26	Fe	92.78	1.17	55.847	78	Pt	131.62	1.3	195.08
27	Co	94.07	1.16	58.9332	79	Au	129.60	1.34	196.9665
28	Ni	94.56	1.15	58.69	80	Hg	121.42	1.49	200.59
29	Cu	95.16	1.17	63.546	81	Tl	123.78	1.48	204.3933
30	Zn	92.16	1.25	65.39	82	Pb	124.14	1.47	207.2
31	Ga	92.86	1.26	39.723	83	Bi	125.52	1.46	208.9804
32	Ge	95.89	1.22	72.61	84	Po	126.35	1.46	209
33	As	98.06	1.20	74.9216	85	At	127.75	1.45	210
34	Sc	100.81	1.17	78.96	86	Rn	97.87	2.2	222
35	Br	103.69	1.14	79.904	87	Fr	87.12	2.7	223
36	Kr	106.06	1.12	83.8	88	Ra	87.12	2.2	226
37	Rb	73.19	2.16	85.4678	89	Ac	111.77	2.0	227
38	Sr	81.11	1.91	87.62	90	Th	123.15	1.65	232.0381
39	Y	90.31	1.62	88.9059	91	Pa	124.48	1.69	231.0359
40	Zr	97.34	1.45	91.224	92	U	136.76	1.42	238.0289
41	Nb	102.90	1.34	92.9064	93	Np	137.58	1.413	237
42	Mo	105.53	1.30	95.94	94	Pu	138.34	1.407	244
43	Tc	107.75	1.27	(98)	95	Am	129.90	1.553	243
44	Ru	109.49	1.25	101.07	96	Cm	139.07	1.407	247
45	Rh	110.15	1.25	102.9055	97	Bk	139.90	1.40	247
46	Pd	109.16	1.28	106.42	98	Cf	147.06	1.304	251
47	Ag	106.66	1.34	107.8682	99	Es	141.99	1.38	252
48	Cd	100.76	1.48	112.411	100	Fm	142.30	1.381	257
49	In	103.44	1.44	114.82	101	Md	142.73	1.38	258
50	Sn	106.25	1.40	118.71	102	No	137.90	1.459	259
51	Sb	107.19	1.40	121.75	103	Lw	144.38	1.367	260
52	Tc	110.15	1.36	127.6		Unq	150.98	1.285	261

The distribution of radii versus 1s atomic attributes is for use by the reader in understanding the equations. Values beyond the named atoms were estimated. Each orbital of an element has an N value that is some distance greater than its N 1s value according to the matrix of that element. You are given the N value of the 1s orbital of each atom. You must define the correct matrix and then devise a spreadsheet to calculate the radius of the atom of interest. The radial value found must be doubled unless the orbital is only partially filled. I prorate the orbital for occupancy before doubling. The radii are quoted in angstroms.

The next Table 3-2 is a listing of the spectra of hydrogen as they enter and their energy is imparted to an electron. It is found in the Appendix. The electron undergoes a reduction of radius as it increases in velocity. The series of energy increases works with first acquiring either the low or high energies first. I deemed the former energies were more likely because greater energies seemed to be required at ionization; and, the table is so ordered. This raises the question that, in an ensemble, any available photon or energy segment might find a home in whatever atom was available. No answer to this is known.

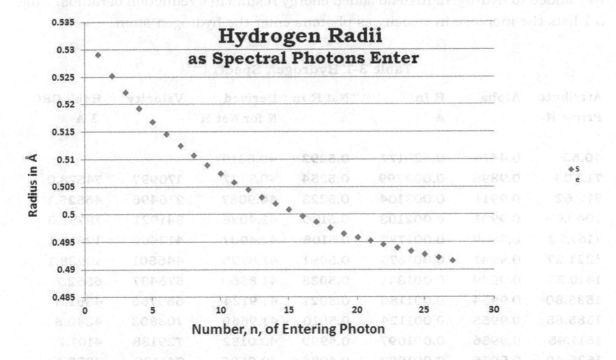

Figure 3.2 indicates that the radius of hydrogen decreases as spectral photons enter. One might expect the eccentricity to become greater not less as the radius decreases. The first figure does not show but only implicates a spiral nature of the hydrogen orbit. The actual difference in radius between the start and end is only 5.29×10^{-11} meters.

Figure 3-3 depicts the eccentricity of the circular Bohr-like orbits for hydrogen. The eccentricity was calculated by taking the radius at half an attribute above and below the circular central radius and then placing these values in equation 3-10. There is a reduction in the eccentricities of hydrogen orbit as successive spectra enter. For some unknown reason the curve is not entirely smooth in the later entries. I had expected that the values would be closer to unity; but, the method of taking equal half-attribute intervals of N yielded values of eccentricity generally much less than a half. However the slope is steep. One would expect the eccentricity to decline more or less smoothly as the total energy of the electron increases when photons are added; but, for no apparent reason there is a second smooth tail to the curve of eccentricity for hydrogen

Table 3-1 delineates the change in radius as successive photons enter hydrogen. The table for helium is much longer and can be found in the Appendix. When the photons are absorbed into the electron the radius decreases which is a consequence of the concomitant increase in velocity. I had earlier found that successive orbitals of atoms of the atomic table had orbital radii that summed to the known radius of that atom, then there should be a similar radial addition that should occur as energy was added to hydrogen. Instead added energy results in a reduction of radius. Table 3.1 lists the increase in velocity as photons enter the hydrogen atom.

Table 3-1 Hydrogen Spectra

Attribute Prime N	Alpha	R in Å	Net R in Å	Derived N for Net R	Velocity	Hdbk CRC λ Å
40.53	0.4414	0.529177	0.5292	40.5319		
712.03	0.9898	0.003799	0.5254	40.7427	170997	74578.0
812.62	0.9911	0.003104	0.5223	40.9087	216496	46525.1
1049.09	0.9931	0.002103	0.5126	41.4026	341021	18751.0
1167.82	0.9939	0.001787	0.5108	41.4907	412460	12818.1
1221.27	0.9941	0.001670	0.5091	41.5725	446501	10938.1
1410.83	0.9949	0.001341	0.5033	41.8560	576437	6562.7
1535.80	0.9954	0.001180	0.5021	41.9124	669756	4861.3
1585.85	0.9955	0.001124	0.5010	41.9660	708803	4340.5
1611.45	0.9956	0.001097	0.4999	42.0182	729138	4101.7
1626.40	0.9956	0.001082	0.4988	42.0695	741130	3970.1
1635.92	0.9956	0.001072	0.4977	42.1203	748811	3889.0
1642.37	0.9957	0.001066	0.4966	42.1707	754031	3835.4
2274.77	0.9969	0.000651	0.4960	42.2014	1339323	1215.7
2274.77	0.9969	0.000651	0.4953	42.2321	1339326	1215.7
2387.18	0.9970	0.000606	0.4947	42.2607	1458072	1025.7
2423.54	0.9971	0.000592	0.4941	42.2886	1497410	972.5
2439.92	0.9971	0.000586	0.4935	42.3161	1515273	949.7

2448.70	0.9971	0.000583	0.4930	42.3435	1524888	937.8
2453.96	0.9971	0.000581	0.4924	42.3708	1530657	930.7
2457.35	0.9971	0.000580	0.4918	42.3981	1534389	926.2

I also give the helium spectra and it will have 126 entries and it is found in the Appendix. As expected there are differences between that of hydrogen and helium. The interesting feature that occurs is that the two electrons both must move at the same time until first ionization occurs. If you let one electron set while the other ionizes the radius of the ensemble is too nebulous to depict. There are two distinct breaks in continuity that can be considered. I have iterated the attributes of the 126 photons several times to cover now the best scenario that I think I can find. It takes eight hours to iterate radial changes to 8 decimal percentage points of the energies of all the photons of the helium spectra. Radii were unheard of as being is coordination with spectra heretofore.

Figure 3.3 describes the progress of helium spectral entries as they finally result in the first ionization. The circular orbital velocity has two changes either side to indicate the swing of one or the other of its two electrons toward the ap-atomic or peri-atomic velocity.

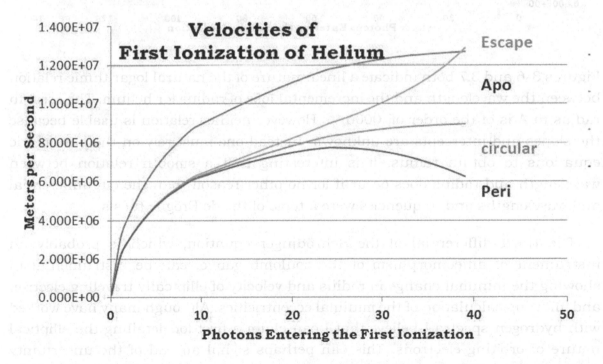

Figure 3-4 deals only with the first ionization of helium. One can see the limits imposed by the escape velocity on the circlar orbit. After the electron having the greater velocity reached the escape velocity I presumed that only the second lower orbiting electron would receive any of the spectral energy. Figure 3-5 then is similar to the first ionization; but, it starts at greater velocity. The second electron as it goes from high to low and back to high radius will eventually exceed escapre velocity as photons enter and it comes to the excape velocity of the second ionization.

$$e = \frac{\sqrt{a^2 - p^2}}{a} \qquad\qquad 3\text{-}10$$

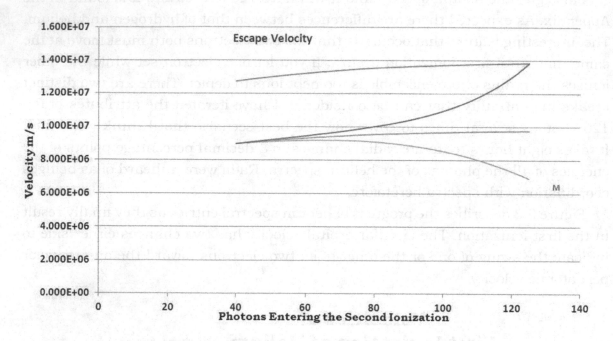

Figures 3-6 and 3.7 both indicate a linear nature of the natural logarithmic relation between the wavelength and the incremental loss of radius for helium. The error in radius in Å is of the order of .0006 %. However neither relation is usable because the slopes and intercepts are unknown. Instead one must rely on the parametric equations to obtain radius. It is interesting that a smooth relation between wavelength and radius does occur if for no other reason than the circumferential and wavelengths and frequencies were a topic of the de Broglie thesis.

The third differential of the Schrödinger equation, which is probably an instrument of diffeomorphism of the coulomb gauge, can be instrumental in showing the minimal change in radius and velocity of elliptcally travellng electron and allowing calculation of the minimal eccentricities. Although many have worked with hydrogen spectra I believe that I can claim a first for detailing the elliptical nature of orbiting electrons. This can perhaps signal an end of the uncertainty principle. One of the shortcomings expressed was that a circular radius could be faulted because that orbit was certainly not elliptic. Another shortcoming was that my circular radius did not reflect center of mass coordinates. Finding of radii of ellipticity corrects that because we can postulate we are at focii with respect to the center of mass. At any rate it is known that for every elliptical orbit there is a circular orbit having the same frequency.

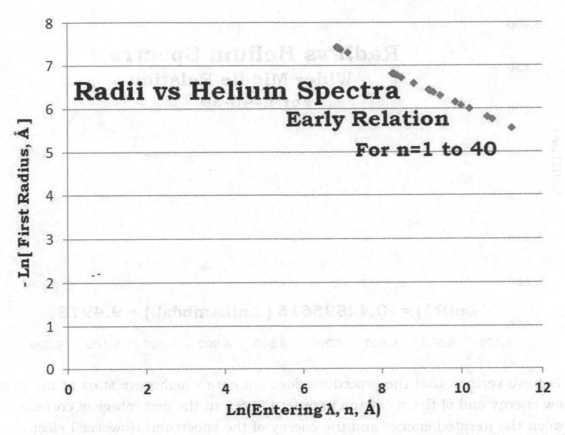

Figures 3-6 and 3-7 deal with the first two portions of the helium spectra acting to create a change in circular radius of the electrons of helium. It is of interest because of a certain amount of regularity that occures in radial change due to the magnitude of the wavelengths of entering spectra. It is not a working relationship because the intercepts and slopes are not predictaable.

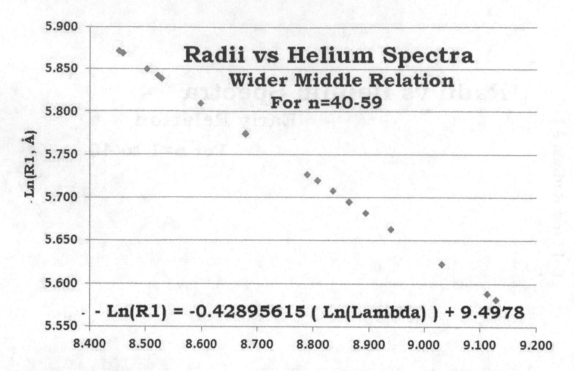

TI have verified that the procedure does not care whether we start at the high or low energy end of the spectrum in order to present the percentage of correlation between the iterated energy and the energy of the spectrum. However I elected to start with the high energy photons first. The iteration does require eight hours. I should state that in the *total cumulative energy* of the entering photons the comparison by iteration is also to six decimal places. A word about the helium presentation is necessary. I had to move both electrons at once until one reached the escape velocity.

If only one knew in advance the slopes and intercepts of these linear relations most of the equations that I have presented would not be necessary. I have designated this set of change in radii with entering velocity as R1 because to obtain the set R2 of final radii the R1 values must be used to diminish the zero point radius of helium of .305 Å. It was assumed that the entering photons must work on both electrons of helium to keep them in elliptic orbit. The elliptic orbit could not be known in advance so a circular orbit from the mean radii of the .3-.31 Å spread quoted in the literature. Knowing the successive changes of R1 it is but a short subtractive step to finding the set of actual radii iterating the The ionizations of helium yields four equations, the middle two have the same slope but only a slightly lesser intercept. This linear technique is not worth much because the slopes and intercepts are not known in advance. It does bolster the idea is some regularity between the wavelength and the change of radius.

To what extent can we trust that the radii and accelerations of the fields for hydrogen and helium found in Tables 3-1 and 3. 2 are valid? Are they reasonable?

I shall recapitulate. First of all the acceleration depends on the unified field times the Poisson. One cannot fault the expressions $E = Mv^2$ and $v^2 = g/r$. Consider the equation below in which M_e, M_P G, K, and Q are all constants.

$$\Delta E = \frac{M_e \left(GM_P + KQ \right) \nabla^2 \left(\phi \right)}{r} = C \left[\frac{\left(\nabla^2 \left(\varphi \right) \right)}{r} \right] \qquad \text{3-10a}$$

There is then a constant C multiplying a function of N, namely $\left[\Delta E = Cf \left(N \right) = C \nabla \left(\phi \right) / r \right]$. This in effect means that the only items left to question are the parametric equations for radius and the Poisson, both connected to my development of some fifty plus years ago. The energy is thus dependent on iterating a value of N. I think that the Bessel equation implies the Poisson without the relative permittivity. Relative permittivity is missing in the Poisson but, is separately found in in the Coulomb constant. I demonstrated that the radial portion of the Schrödinger radial equation, a portion of the main Schrödinger wave equation, was equivalent to the Taylor series and its derivative of my Bessel development when I redefined some variables. I have earlier explained that the series involved in my Bessel development were two infinite Taylor series, one for F and another for its derivative, and expressible in two connected parts as A + B + 8 = 1. They definitely correlated material balance equating output equal to input. After writing A + B + 7 = 0, I determined that A, the vector potential was also the scalar potential being the holdup, as it were. Then I reasoned that B + 7 was always zero because *any value of N* produced zero steady state, even though differing N caused alpha to be variable. The variable alpha changed the value of the two attributes of a second set of radii and thereafter calculating the eccentricities.

Function of N from the B + 7 as the zero time evolution factor I was able to change the A vector form into a scalar quadratic in N with the numerical value of the Poisson. It is the scalar quadratic in N that allows a numerical value to be found which is the Poisson, or electron density. It was because of a prior serendipities event in which I found the connection of alpha and its associated N to the Bohr radius that I was able to conjecture the solution to the Bessel equation in radial coordinates. This has proved itself greatly by allowing iteration of the spectral radii of hydrogen and helium.

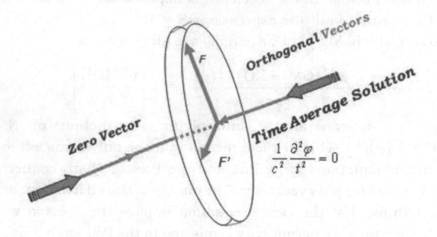

Figure 8 is a depiction of the d' Alembertian usually shown. I have shown the vector portion as two perpendiculars that are orthogonal in a plane perpendicular to the time variable. Perhaps we can call this a spacetime d' Alembertian; but, the thought only lately occurred to me as someone else suggested that this book was about spacetime. There is a relation between the last two terms of equation 3-10 b and or equation 2-1 a:

$$\frac{1}{c^2}\frac{\partial^2 (\phi)}{t}$$

3-10 b

It is required to embellish the gravitational term as well because it can be expressed as an electromagnetic equivalent. In succeeding chapters there are so many areas of chemistry in which there are possible applications that there needs to be many more people who can lend their expertise to future developments. Lest you think that development to this point was easy, recall that the process of development has taken me over fifty years. My local Physic Teacher's Organization has tired of hearing me. The second quantum theory era spawns ninety-three years and is not the mind of one person. The development of the book has occupied my mind for over fifty years and few if any as yet realize its impact on the thinking of the quantum world.

It is reported on the internet that various prominent physicists have despaired ever finding a correlation of the helium spectra. One can always ask whether what I have presented here can be proven. The energy that goes in via the photons has to correlate the mass times velocity squared of the two electrons in helium. In hydrogen each photon imparted energy as velocity increase for a single electron. The radii are correlated by the parametric equations which are correct at each point. The correctness of the parametric equations for radius at the next point depend on the radius of present point that is required to accept the next photon energy determined by the list of spectra. That radius is reached by the parametric equations to the limit of the spectra.

The Poisson is an electron density factor and is a relation that is inherent in the Associated Legendre equation. Figure 3-9 indicates how it is derived from that equation. Since every change in radius is occasioned by a change in N, the zero time energy term, B+7, remains zero as the time like variable N increases. A new input of energy take us to a new time zero radius in going on to ionization. One point emerges from this is that we have jumps indeed with each arrival of energy culminating in a new time zero evolution et cetera. We can claim internal consistency.

$$BF + \frac{2d}{d\alpha}\left(2(1-\alpha)\frac{dF}{d\alpha}\right) = 0 \quad \text{Now let } B = \frac{A}{2}$$

$$AF + \frac{2d}{d\alpha}\left((1-\alpha)\frac{dF}{d\alpha}\right) = 0 \quad \text{Where } F = \frac{1}{1-\alpha}$$

$$AF + \frac{2d}{d\alpha}\left((1-\alpha)F'\right)$$

$$AF + \frac{2d}{d\alpha}\left((F)\right)$$

$AF + 2F'$ allowing A to equal N+5

$\nabla E = (N+5)\,F + 2\,F'$ we have the Poisson

portion of the Attribute Wave

Figure 3-9

It is of great importance to extend the algorithms developed to atoms beyond helium. It is also important to note that no other satisfactory means for extending the methods of calculation of the spectra of hydrogen to helium has been found. Currently no other correlation of radii, velocities and eccentricities exist. Is my method correct? The correctness depends on the veracity of the parametric equations which I have detailed. I have detailed that when one uses the parametric equations to iterate the change in radii with increase of energy then the values of the radii of hydrogen and helium that these parametric equations describe are consistent with the Associated Legendre equation through the time average of the Bessel function and the definition of the Poisson as well. I trust that the methods that I have shown are more nearly correct than any previous methods and are as exact as anyone could want. They are classical and because of this the entire façade of the phase space becomes suspect as far as probability is concerned. I have found that it is extremely difficult to find any text that describes anything done via quantum mechanics to solve a viable problem.

$$\nabla^2 E = (N+5)F + 2F' - (N-28)F^{-1} + 7 = 0 \qquad \text{3-11}$$

$$\nabla(\varphi) = \frac{(9N^2 - 273N + 588)}{49} \qquad \text{3-12}$$

$$\alpha = \frac{(N-35)}{(N-28)} \; ; \; r = \frac{\alpha \, Ln(\alpha)}{\sqrt{N+6}} \quad \text{The parametric equations} \qquad \text{3-13}$$

$$g_{universal} = \frac{(GM + KQ)\nabla\phi}{r^2} \qquad V = \sqrt{rg_{universal}} \qquad \text{3-14 a b}$$

Calcium Half Orbital Radii

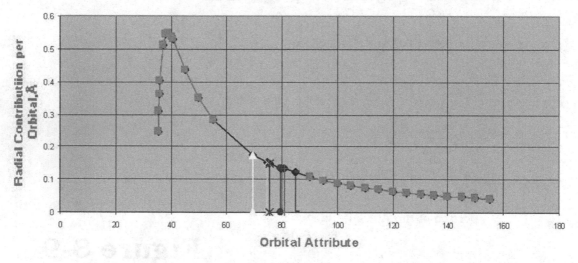

We must define a matrix of attributes, N, distinct for each atom whose N1s index is unique for each element. The N1s index Np is called the principal attribute. The 1s columnar indices increase by fives and replace the principal quantum numbers. The l = 0, 1, 2, 3, 4 quantum numbers add to the Ns orbital values. So, the adjacent s, p, d, and f are each greater than the previous by 1. For calcium the attribute, matrix would be:

69.65948		
74.65948	75.65948	76.65948
79.65948	80.65948	
80.65948		

The N1s real numbered value of N has to be iterated so that when the orbital attributes are subjected to the parametric equations the matric of orbital radii, when doubled, sum to the radius of calcium. For any atom every orbital has an N attribute thus creating a matrix with a unique N1s attribute, and has a similar α matrix, a similar radial matrix, a Poisson matrix, a $g_{universal}$ matrix, and a velocity matrix. These matrices can be calculated via equation 3-13, 3-12, 3-14 a b. That radial value is for one electron and is doubled for full orbitals and is prorated for those orbitals with 6, 10, or 14 electrons. The matrix of radii is summed and doubled to obtain the requisite radius of calcium where the mass of the calcium atom and the charge on the mass of calcium is used in equation 3-14 a b.

It is convenient to spreadsheet the entire series of matrices. For this we will utilize the classical equations that are involved with these electron mass and central mass and charge relationships unique to the atom in question. We noted in chapter one that there is an electron density that can affect the gravitation constant because gravity can be expressed in terms of an electro- magnetic function Figure 3-11 shows the usual Periodicity of Radii with Atomic Number above. This lower matching periodicity is an inverse of the one above. We also noted in that chapter that because of the radical unity in the rearrangement of the Rydberg equation it is possible to anticipate the unified field as a sum of the gravimetric and electric centers of the atom. Masses exert a gravitational effect that is expressible in electric terms so that the Poisson.

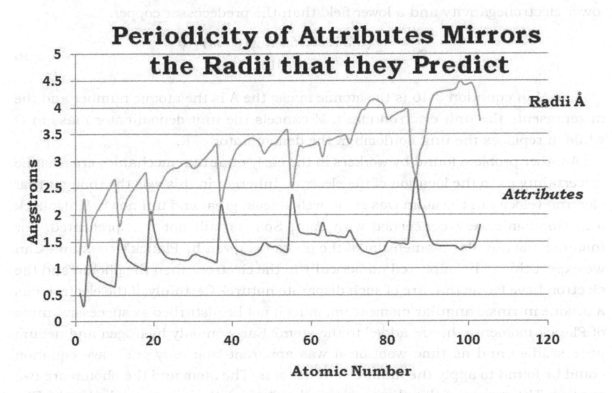

Obviously in order to use the Poisson in a field expression one need not supply the permittivity because it is inherent in the Coulomb's law constant which was already in place. It is possible to represent the calcium radius by the Lesbegue integral:

$$R_{Ca} = 2 \sum_{N_P} \sum^{N} \frac{\alpha Ln(\alpha)}{\sqrt{N+6}}$$ 3-15

This is depicted in the Figure 10. The periodicity of the radii with atomic number is well known. What is unique is that there is also a periodicity of the iterated principal attributes, Np, with atomic number.

There is some indication that the Poisson can be used as an index of the electronegativity. The natural logarithm of the unified field is also a possibility. For some atoms the question is should we use the outer orbitals, the outmost orbital or the average of some or all. One of the difficulties that quantum mechanics encountered was in_assessing the electronic densities within atoms in order to apply use of normal conventional classical procedures. The Poisson could handles this with a minimum of difficulty. The match is not necessarily so good as to be useful. It would seem that if there is value in knowing the electronegativity of the unified field it would be in some relation to the Poisson. The anomalous behavior is in the lower portions of the series. The magnetic materials all have nearly the same electronegativity of 1.8; but, there is a spread in unified field Poisson. Zinc has a lower electronegativity and a lower field than the predecessor copper.

$$g_{Unified} = \frac{(GM1836.15 + K\,A(Q/m)\nabla^2\phi}{((1e-10)r\mathring{A})^2}$$ 3-16

The M in equation 3-16 is the atomic mass; the A is the atomic number and the m represents the unit electron mass. M cancels the unit denominator mass in G while m replaces the unit coulomb in the denominator of K.

Another problem found by workers in the early quantum mechanics era was the uncertainty as to the location of the electron. Inherent in this was the thought that electron velocity in the atom was great, well at least great and unknown. Uhlenbeck and Goudsmit were concerned with spin. Spin is still not well presented. For instance the angular momentum of the photon is given by Planck's constant. Can we expect this to be imparted numerically to the electron when the photon and the electron have forms that are of such disparate nature? Certainly, if the electron has a unique intrinsic angular momentum, must it not be disturbed as successive units of Planck momenta, h, are added to the atom? Early on only hydrogen and helium were studied and as time went on it was apparent that only one wave equation could be found to apply the multi-electron atoms. The atom and the photon are two entities. The energy of the photon enters the atom as the energy implied in *hf*. The photon having no mass can have no momentum to transfer.

My notion of spin is characterized by narrowing the somewhat uncertain radial location of the electron due to the fact that it is in an elliptical rather than circular orbit. It is known that for every elliptical orbit there is a circular orbit with the same frequency. To this end we note that the radii given by the parametric equations cause the electron to circulate about the nucleus of the atom. When the eccentric distances are found the nucleus becomes the focus of the two disparate radii. If they depart in the atom from circularity due to their assuming an elliptical orbit the amount of deviation is coded by no more than a half an attribute in magnitude else there would be a clash with electrons of orbits above or below. An earlier criticism of the orbit was that it did not account for the center of mass coordinates is minor if you can specify the two distances of interest of the ellipse. Now let us reduce that uncertainty somewhat. If you have major and minor distances from a location you have an ellipse. If you consider two spins about poles perpendicular to electron orbit you can postulate that the two electrons thus spinning cause a magnetic reversal through the plane of electron orbit every half revolution of the electron about the nucleus. As the radius changes from high radius or N-½ to low radius or N+½ one goes from Ap-atom to Peri-atom generating an ellipse. This slow magnetic reversal generates the ellipse; and, this elliptical path can accommodate two electrons. There is another spin besides the magnetic and orbital spin making a total of three spins. The third spin is about the magnetic spinning polar axis. Of course every elliptical path contains but two electrons. What happens when an orbital has five or seven elliptical paths? This is why the p orbital probably cannot be as it usually is shown. One can stagger three, five, or seven poles as one, two, or three poles either side of a central pole with electrons staggered so as not to clash at crossing points. All electrons in an orbital have the same N value so this seems possible.

What exactly is the nature of the word quantum as it appears in quantum mechanics and the new mechanics of attributes. The unit of one quantum is a photon whose one similitude is not a constant angular momentum but its unique energy. Energy to warm is not questioned in a spectrum tube. Energy to effect a structural difference is the question. Why must a given atom in a given state only to suffer change if struck by a photon of a given amount of energy. I do not believe that this is true of orbiting spacecraft so why should it be true of electrons orbiting atoms? It appears that it has been tacitly assumed that angular momentum has to be conserved and resides in the electron that at this time can only accept the wavelength of the photon to complete the orbit. There is no such caveat inherent in the unified field! There is no guarantee in the action of voltage of the spectrum tube! Electrons all are taken to be the same mass. Is this why we cannot find an integral quantum jump in the solar system? What convenience is there in quantizing anything? Each entering photon carries a different amount of energy which has been somehow connected to the n or quantum number. The energy carried or emitted by a value of the Lyman series is not the same as one of the Balmer series. Why must any electron struck by just any photon not move? *In short there is evidence that a substance could take up a large number of the same package of energy unless it is a laser!* Since we have no way of knowing which atom of an ensemble is emitting or being hit by a photon there is no basis for the numerical certainty we call quantization. Each entering photon is really a change of one albeit countable

This brings us to grips with the idea of quantum gravity. We saw in chapter one that by using the idea of a macroscopic quantum number in the de Broglie relation that we could find a quantum number for the position of the planets. However, this was a real, not an integral number. There is no calculable quantum number n that is required in attribute mechanics that can move a planet. We have and we must conclude that there is no integral bar to the *classical nature of attribute mechanics and general relativity despite the fact that the Lagrangian was used to formulate general relativity and the Laplacian can be found from the radial portion of Schrödinger's equation.* There is no need to try for a quantum gravity that is a requisite part of general relativity. Recall that the only thing about a quantum of energy that is the same is its use of Planck's constant which only has the units of angular momentum. We have tested hydrogen and helium for angular momentum and found that only a few percent of that implied momentum *h* of the entering photon is found. There is no integral multiple of *h* in the calculation of the radii, the Poisson, or the unified field.

Let us address another aspect of our classical atomic solution. We apparently have forgotten that there is expansion of materials with temperature. We have assumed room temperature and atmospheric pressure for our attributes of the elements. When in a future chapter we will deal with the properties of the atoms and have to contend with elevated temperatures then we will have to define a change in attribute with temperature.

I have explained how to define the aufbau matrix of attribute numbers N which is the argument for alpha. Alpha and N are the arguments for radius. N is the argument for the Poisson, namely, $\nabla^2\phi = (9N^2 - 273N + 599)/49$. The unified field, and the Poisson are arguments for velocity, $v = \sqrt{\frac{(GM + KQ)\nabla\phi}{r}}$. The 1s value of N is unique for each atom. It is called the principal attribute and is found for all atoms of the periodic table in the Appendix. You can set up your orbitals for a given element and check the values of given 1s principal attributes. It is sometimes difficult to set up the original matrix of a heavy atom and especially to decide how to prorate the outer orbital; but, the value of radius of an atom comes from a doubling of the filled orbitals and adding the prorated unfilled orbital values of r. There are several ways that this can be spread-sheeted. The fastest way is to work downward on the sheet and calculate the radius by a single summation. In spread-sheeting with spectra work across the spreadsheet.

Chapter Four

Attribute Mechanics and the Temperature Shift

It is not enough that the radii of hydrogen and helium have been now so carefully determined in a novel manner. Considering the nature of an ensemble that is present in a spectrum tube, the solution is totally complete. Not only can we expect a range of states since many frequencies are being emitted but also there is an average temperature of the mixture to be considered. This will alter radii slightly. If one assumes all possible states are emitting, then there are some 128 different, such possible radial states in an ensemble of helium. The probability of any one state is only somewhat calculable if one takes note of the range of frequencies emitted. Measuring the frequencies of a given ensemble may not find the amounts of those near ionization nor satisfy the energy balance. The intensities are needed as well. We might arrive at a credible answer if the intensity is directly related to the frequency. The amount of current and its frequency is also going to be a factor. Still the fact that each state is most certainly at a temperature of its own making constitutes yet another variable that is operating on the "average radius" of the atomic radius. In the previous chapter I assumed that the ensemble could be taken to its limit. Helium, in particular, can rise to a temperature of 5000 °F during excitation. We can calculate the effect of the attribute on the radial span differential using the parametric equations and find the change in attribute with either a change in radius or degree. This is of course a number that differs for each element.

Because gas molecules are so far apart compared to liquids and solids it is not possible to use volumetric data much less radial distances based on the perfect gas law in equation 4-1. We did find the radius of a monatomic ion at ionization in the case of hydrogen and helium. In so doing we easily got the change in attribute from the energy added without knowing a temperature. Measurement of the change in volume with temperature will allow finding the equation of change in radius with temperature. However this is without adding any energy to alter the radius by uptake of spectra. Finding spectra of atoms such as chlorine may also be done. A similar technique for solids such as $CaCl_2$, where ionization spectra are in the literature, then the change in radius with change in energy can be calculated. We will not

67

be able to find a correction for principal attributes for gases but using volumetric coefficients or linear expansion coefficients of expansion we may characterize those elements that are solids from 2s to Uranium solely for temperature change without spectral energies being added.

$$PV = nRT$$ 4-1

A novel method of finding some bond energies may work. The basal energies of the solid elements may be found. This can be repeated if electrons inserted as they would be if the two elements had reacted. Sub traction of the two unreacted from the two reacted and compared with literature values of bond energy may just be close the correct measured value. If not tweaking of the method of assigning values of energy to unfilled orbitals may bring the method into compliance.

There are numerous descriptions in the literature concerning the state of the art with regard to quantum mechanics and general relativity.[10] In part one of his two part article, Goldstein comments on the points of view of some of early 1900's architects concerning the quantum theory. He reports on one current way to reconcile this hiatus that stems from the attempts to reconcile this new quantum mechanics with general relativity. This concerns the inability to scale from micro-atomic to solar proportions, where the mechanisms used by quantum mechanics now seem to be most useful in doing kinetics where probabilities usually play an important.

I believe that we have already reached the "holy grail" of physics namely the reconciliation of attribute mechanics with general relativity. The scale up from atomic proportions to solar proportions is delineated. The rearrangement of the Rydberg equation and the definition of test particles suffice to delineate scale up. Recall that in such a reconciliation quantum mechanics would suffer the most change according to Roger Penrose. Note the smooth transition of the real numbered quantum "jumps" of the planets in Figure 4-1

---

[10] Goldstein, Sheldon, Quantum Theory without Observers, Part One, *Physics Today*, Mar. 98, p. 42.

Figure 4.1
Variation of Quantum Number

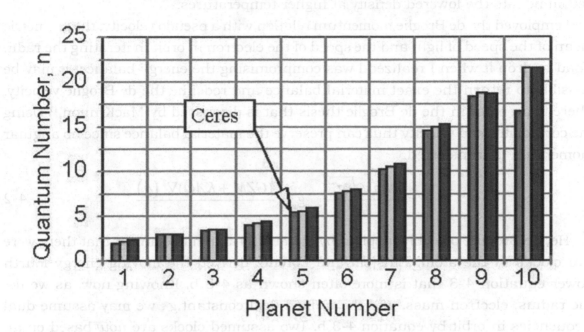

We need to find a way in which, for a given higher temperature, that we can find a *lower*, attribute, which would anticipate the increase in volume of the solid material as the temperature rises. We know how to find an increase in radius with lowered attribute numbers. We can find the variation of attribute of an element with temperature using volumetric expansion data. This may prove to be a step to determining an entropy.

The exact amount of change in radius will depend on the material, the temperature change of course and that dr/dN can be calculated by the parametric equations. We assumed that the temperatures given range from 68 to 5050 °F.

One of the disconcerting facets of quantum mechanics is its asserted inability calculate radial distance. Instead of atomic radii we are left with probabilities. Yet from the associated Legendre equation of the Schrödinger wave equation, the Bessel function and the parametric equations, we have been able to construct the radii of all atoms. We have defined a viable unification of both atomic and solar fields. The solar, and cosmic, gravitational fields can be expressed in electromagnetic terms in place of Newton's Universal Gravitational Constant so that the Poisson is apropos. The Poisson does not have a cosmic value; but, it will depend on electrons ejected on orbiting matter. Because our *radial solution* is consistent with the radial portion of the Schrödinger equation posed by the associated Legendre equation, then I can claim that my numerical attribute radius is a splinter off of that wave equation which does not embrace the Lagrangian portion which renders the radial solution classical. This may allow some consolation to Quantum Mechanics as to a possible claim to reconciliation to general relativity since they both employ the Lagrangian. The numerical attribute will decrease to reflect the requisite greater radii that accommodate the lowered density at higher temperatures.

I employed the de Broglie momentum relation with a pseudo velocity, the geometric mean of the speed of light and the speed of the electron in orbit in iterating the radii. I had to drop it when I realized I was compromising the energy balance. It may be possible to return the exact material balance and redefine the de Broglie velocity. There is an error in the de Broglie thesis that is discussed by MacKinnon.[11] Using the correct iterated velocity thus can preserve the material balance since no angular momentum is transferred.

$$f = \frac{2\pi m_e c \sqrt{r g_u}}{h} \qquad g_u = \frac{(GZM + KAQ)\nabla^2(\phi)}{r^2} \qquad \text{4-2}$$

He, de Broglie, possibly inspired by equation 4-3 a, had claimed that there were two clocks in the atom. Einstein and Lorentz derived the moving energy fourth power equation 4-3 that is more often shown as 4-3 b. Knowing now, as we do, the radius, electron mass, and the unified field constant,g_u, we may assume dual frequencies in orbit by equation 4-3 b. Two assumed clocks are *now* based on an electron velocity and the speed of light. The variable c v is the square of a pseudo de Broglie velocity. Now we know the electron velocity, which was unknown then. They would have had no way of knowing that energy and angular momentum balances conflicted. Dividing $m_o^2 c^2$ of equation 4-3 b by its equivalent $h^2 f^2$ one arrives at $m^2 cv$ equal to zero. We cannot now construe the pseudo de Broglie relation to imbue the system with two clocks or angular momentum because photon m_o does not mover. Not so fast! There is only one substance in 4-3 a, b.

$$m = \frac{m_o}{\sqrt{1 - \dfrac{v^2}{c^2}}} \qquad \text{4-3 a}$$

$$E^2 = m^2 c^2 v^2 + m_o^2 c^4 \qquad \text{4-3 b}$$

It is erroneous to assume that the two masses of equation 4-3 b exist at the same time. The subscripted mass goes relativistic and cannot be the mass of the electron. One cannot extend the mass of the photon to be the electron mass and demand a momentum of $m_e \sqrt{cv} \, r$. As with any energy in orbit there is an angular momentum. The photon has no mass; and, it has only energy to transfer.

[11] Mackinnon, Edward, de Broglie's thesis: A Critical Retrospective, *Am J Physics, Vol. 44, No. 11 Nov 1976, p. 1047*

I earlier defined an absolute pseudo momentum involving $m_e\sqrt{cv}\,r$. and thereby defined the "two clocks of de Broglie" by involving the velocity of the electron and the speed of light. I iterated a radius indicative of an energy $m_e cv$ as being equal to photon energy $h\,f$. *Obviously the calculated* $m_e v^2$ required a different radius in order to be in balance. Perhaps this is one of the difficulties that caused Bohr's notion of complementarity to arise. Pauli declared that if the use of one classical concept excludes knowledge of another, then both concepts (e.g., position and momentum) are complementary to each other. While Schrödinger's equation builds Fourier frequencies attendant with multiple modes of variable probabilities we employ possibly *similar variables* that could be functionally connected to probabilities. This analogy is not exact because we can render null Paull's idea of complementary variables. We have a set of frequencies that equals $v/2\pi r$.

Generally, the electron velocities in atoms are essentially non-relativistic, although; they definitely are in sub atomic quarks. Since electrons are smaller mass anti-protons I would like to call one of their three central positive charges a down quarkino. The mass of a tau electron results from six relativistic orbiting negative charges each q/3. Electrons, protons, and neutrons can be described in terms of relativistic orbiting sub-particles. A relativistic particle does not appear to be gravitationally attracted.[12] Rather, in the atom, unless highly excited, orbital velocities are more nearly of the order of the speed of sound. Even the geometric mean of this speed and the speed of light is not relativistic.

Partially filled orbitals have electron radii that are pro-rated for occupancy. If the orbital is partially filled its defining electrons are used to prorate the radii during summation. Just how to configure the outer orbital is sometimes a problem which may alter things slightly. For this reason it is best if you just treat the 1s values of the principal radius found in one of the Tables of the Appendix as initial points to learn how use the calculation. Exact values of published radii by your version of the aufbau matrix may require a slightly different N1s attribute should you alter the matrix for summation. Alteration of the attribute matrix inadvertently can occur. Next we must consider how equation 4-4 a below is related to Schrödinger's equation. The Schrödinger radius described as the variable ρ has not received much attention. Thus α is the variable in which the *Schrödinger radial variable* must be inverted in order to acquire the true radius. This latter definition of inversion is often used in potential theory.[13]

$$r_i = \frac{\alpha Ln(\alpha)}{\sqrt{N+6}} = \frac{\alpha}{\rho} \qquad\qquad 4\text{-}4\text{ a}$$

[12] Eisberg, R.M., Fundamentals of Modern Physics, John Wiley & Sons, New York, © 1962, p. 30 f.

[13] Eves, Howard, Survey of Geometry, Volume 1, Allyn & Bacon, Boston, © 1963, p.145f.

The variable rho has not had a role as yet because there is scant or no necessity to present it. *Rho* was accorded to be the variable designated to represent *radius* in the associated Legendre equation. I found it necessary to *invert rho in alpha* in order to express the Bohr radius which was one of the serendipities. The variable α is thus a necessary adjunct to the radial portion of the Schrödinger wave equation. It has values ranging from .38 to 1.0 at low to high values of N. At or below an N value of 35 a radius is not defined. The Bohr radial value of .529 Å is achieved at a value of N circa 40.54 which is just below the maximum of the distribution. It is found again on the other leg of the distribution of radii. No use has been found for attributes left of the maximum position of the Poisson Distribution Curve. This distribution of radii with attribute, N, is skewed right and looks more as if it were a Maxwellian Distribution Curve.

Therefore ρ is a measure of a normal (pseudo radial) solution to the wave equation in terms of the Schrödinger development. Under auspices of the attribute equation ρ is also variable ostensibly used to represent the radial solution to the radial separable portion of the Schrödinger equation. The true Bohr radial value is achieved by inversion of α in ρ so that the atomic case is an inverted potential. Einstein used an inverted Newtonian potential to define the precession of Mercury and the bending of light by the mass of the sun.[14] It is a known fact that frequencies are preserved under inversion. Consider that the first two terms of the Bessel wave equation constitute an orthogonal pair. Refer now to Figure 3-1. The remainder of the Bessel equation defines α in terms of N and has the role of time dependency equated to zero. There are three well known equations that involve the numeric values of the Bessel equation. The overall equation is the d' Alembertian which is useful in relating or defining the vector and the scalar potentials in electromagnetic theory.

$\nabla^2(\phi) = \nabla^2(\phi) + \frac{1}{c^2}\frac{\partial\phi}{\partial t^2}$ Where ϕ is a harmonic function $\qquad$ 4-4 b

The last two terms are the time evolution terms, $\frac{1}{c^2}\frac{\partial\phi}{\partial t^2}$ which is zero for any N.

$\nabla^2(\phi) = 0$ In the absence of charge q is the Laplacian. $\qquad$ 4-4 c

See the vector potential in figure 3-1, which is zero in N direction.

$$\nabla^2(\phi) = \frac{4\pi\,Poisson}{4\pi\varepsilon_0} = [KQ(9N^2 - 273N + 588)]$$

$\qquad$ 4-4 d

When Q, is present the Poisson is represented as a Quadratic in the attribute N.

---

[14] Eves, Howard, loc. cit.

Equation 4-4 b is essentially a d'Alembertian. The first two terms are the vector potential which can be converted to the scalar charge potential because $F^{-1} = 1 - \alpha$ which allows alpha to be calculated in terms of the attribute N in equation 4-4 f. Thus $a = (N-35)/(N-28)$ and this is half of the parametric equations. You now can substitute for F' and then for α in 4-4 e. This results in $\alpha = f(N)$. Substitute for α in the first two terms of 4-4 e and solve for the quadratic of 4-4 d. Note that the derivative of F with respect to α is really F^1.

$$\nabla^2 \phi = (N+5)F + 2F' - (N-28)F^{-1} + 7 = 0 \qquad \text{4-4 e}$$

$$\frac{1}{c^2}\frac{\partial \phi}{\partial t^2} = -(N-28)F^{-1} + 7 = 0 \qquad \text{4-4 f}$$

I have not found it necessary to find the function ϕ. One can imagine time to be orthogonal to the first two terms constituting the Poisson at steady state. One finds alpha in terms of the attribute N in the last two terms of the Bessel function which is zero because these last two terms are the time average terms. Introducing the definition of alpha in terms of N into the vector function of the first two terms yields a quadratic in N which is the scalar potential, a quadratic in N which was used in chapter three to obtain eccentricities. It is known that the scalar and vector potentials are the same in equations that are in cylindrical coordinates, which the Bessel equation is.

So, the real scalar value of first two terms constitutes the usable Poisson, since they are orthogonal to and zero in the direction of the third term. Since the first two terms are vectors the represent the vector potential. That they can be expressed as a scalar function of N they represent the scalar potential. For Bessel functions in cylindrical coordinates the Vector and scalar function have the same value.

This Bessel equation is a cylinder function and yields radii as a function of N as equation 4-4 indicates. We increase the N value to add a distance from the previous orbital. The very complicated diagrams seen for p and especially the d and f orbitals can thus be greatly simplified and calculated because the poles of each two-electron sub orbital lines of the overall d or f orbital is unique. In other words even the p orbital usually shown with one pole orthogonal to the other two must be tied to one overall pole with the others. How can the electrons keep from crashing? Two electrons can lie in each orbital pole staggered 20 to 30 degrees from the central orbital pole. For a d orbital there are five sub-orbitals with two poles either side of a central pole with electrons that can be appropriately scattered so as not to clash.

These radial values define time average orbits roughly on a sphere in which the mutual repulsion of the electrons on the sphere occurs through the Poisson. No further adjustment other than the principal attribute and the attribute matrices are needed in going to succeeding atoms because the fit is to an accepted value of radius for that atom. One of the unsolved problems preventing the calculation of orbits was that there was no good way to assess the repulsion of the electrons. The Poisson solves the problem handily.

It is known that for every elliptical orbit there is a circular orbit having the same energy. Since the Poisson is a quadratic function of N we can perturb the Poisson in the amount of its derivative or else simply perturb the N values by $\pm$ ½ ΔN which is equivalent to $\pm$ ½ n and find radial values which will define an elliptical orbit. The problem is to find the shift of the attribute with temperature. This is perhaps most profitably done for the transition solids and rare earths because they are compounded now in the solar energy field.

The *attributes are not integers*. Shell and sub-shell position numbers are spaced with regard to integral differences in N, the attribute. Recall that the quantum numbers of the planets are not integers either. The values of N allow us to show that the plot of atomic radii is the mirror image of the plot of radii versus atomic number. The spin results from a change in radius occurring as a $\pm$ ½ ΔN from the time average attribute position. The variable n of course is a consequence of the Rydberg equation and spawned the concept of the quantum jump so vilified by Schrödinger. The jumps are more probably nearly smooth transitions as energy arrives from the photon, a fact that would please Emil Schrödinger. The old *principal quantum number* jump, n = 1 to n = 2, becomes now a five-integer difference in attribute N1s = x to N2s = x+5 between shells and one-integer difference in attribute between sub-shells p, d, and f. In practice a non-integral or fractional movement can occur, as the result of a temperature effect, because we know that temperature causes radius or volume to increase. The entire set of electrons must shift with temperature. The takeoff point in one orbit and the landing point in the next surely do not occur at the same radial direction each time. The attribute matrices are constructed from the aufbau concept; however, neither their N values nor their differences are considered to quantum numbers.

This does create wonder as to whether the quantum designation of quantum theory is a misnomer. Recall that the supposed n units of angular momentum inherent in Planck's constant when expressing the entering energy, *h*f, of the entering photons are not conserved. Angular momentum of electrons in orbit may be found easily because the variables of its definition are readily available; but, it is small compared to the angular momenta of multiple Planck constants.

When the field increases, so does the frequency according to equation 4.2. With increasing value of N the electrons are ever closer to their underlying shell. In the early sixties I derived this Bessel equation to solve a problem in solvent extraction for my employer Pfizer. The variable α was a constant partition coefficient of one solvent and was not a function of stage number, N as it is here.[15] Pfizer has the details of its use in separation processes none of which has relevance for the mathematics of

---

[15] Brooks, J.O., New Atomic Periodicity, 1996.

this book. I have indicated that the Poisson portion of the Bessel equation has a one to one correspondence with the Schrödinger equation radial separation equation called the associated Legendre equation.[16]

Before 2000, I had derived matrices of frequencies, bond lengths, and bond strengths from the unified field equation. The method requires types of selectivity which render it questionable. A better method must be found.

Those who might question whether Maxwell's equation applies to mass as well as to charge should note that radii depend on mass and charge remaining constant. There is a no cosmic equation for radius that does not depend in some fashion on Newton's law. This is why in a later chapter we may examine "dark matter" in terms of the electronic dependence of Newton's law, on local electron densities as shown in chapter one. Dark energy is possibly the result of many electrons landing on the orbiting mass. Indeed the concept of "dark energy" can also possibly be an artifact of the delusion that the velocity term in Einstein's general relativity equation is outwardly directed. Instead velocity is a rotating tangential vector of the hypothetical velocity needed to orbit the source at the distance a photon might be from the source.

Quantum gravity is the name for reconciliation of general relativity and quantum mechanics. It was not thought of in those terms at its inception. As a *real* planetary quantum number it may have something to do with precession. The concept of detecting a graviton is probably remote. An application of the parametric equations to solar radii is found given a revised radial constant as in equation 4-6. Inversion is necessary because radius increases with respect to increase in attribute number. Recall equation 4-5 the one remnant of the Rydberg rearrangement is the equation that gives us entrance to solar phenomena.

$$G = \frac{nhv}{2\pi} \qquad \qquad 4\text{-}5$$

Equation 4-6 fitting the change in the constant N for solar radii that I suggested in a paper, copyrighted in 1996, does not correspond to the N values found by the de Broglie relation.[17]

$$R(N) = \frac{-10c\sqrt{\frac{2}{3}}}{\alpha Ln(\alpha)} \sqrt{N+6} \qquad \qquad 4\text{-}6$$

The values of N in Table 4-1 below (shown in Figure 4.1) were revised in chapter one to reflect the changed nature of n in the solar system due to the de Broglie relation *inherent* in equation 4-5.

---

[16] Pauli, W., Theory of Relativity, Dover, New York, © 1958, p. 202.

[17] Brooks, J.O., Quantum Mechanics and Attributes: Complete or Not, © 1996. p. 19.

Einstein's equation is definitely harder for one to understand than Friedman's version of it in three dimensions. It still employs Newton's gravitational constant, G. No one has tried finding attributes, N, for galactic systems. Feynman derived Maxwell's equations from Newton's law.[18] Feynman's work was published later by Freeman Dyson because Feynman submitted the result to Dyson saying that he did not believe it. Richard J. Hughes explains the nature of Feynman's proof saying that the Lorentz force law and two of Maxwell's equations can be deduced from from the minimal assumptions about the equation of motion and quantizations of non-relativistic velocity

Better quantum Numbers are given in chapter seven when velocities used are the mean solar velocities and the equation $n\,h\,V = q\,G$ is used.

Table 4.1

Planetary	Quantum N	Numbers Velocity m/sec	and Constants RM x 10^{-11}
Mercury	2.12	47734	0.59
Venus	2.90	34889	1.08
Earth	3.42	29657	1.59
Mars	4.21	24052	2.28
Jupiter	7.79	13017	7.78
Saturn	10.55	9611	14.26
Uranus	14.95	6778	28.69
Neptune	18.72	5415	44.95
Pluto	21.49	4716	59

particles.[19] A version of this result is found to hold in classical mechanics and it is shown that, instead of limiting interactions to electromagnetic interactions, Feynman's derivation is a rederivation of the *constraints* on the velocity-dependent generalized forces that can be accomodated by Lagrangian formulation. The generality of the result is illustrated by showing that the particle motion in a noninertial frame or in a weak gravitational field also satisfies the constraints. That the Lorentz covariant version of Maxwell equations admits an obvious n-dimensional generalization is the observation of the Russian Z. K. Silagadze of the Budker Institute of Nuclear Physics. Perhaps a more explicit generalization would entail noting that the Lagrangian simplifies to Newton's law when the distance is radial. He undoubtedly made an assumption. This assumption *had to be that there was some connection between Newton's and Coulomb's law constants*. We found this value in chapter one. We also showed its nature. This essentially proves that gravity is

---

[18] Dyson, F.J., Feynman's Proof of Maxwell's Equations, *Am. J. Phys.*, 58 (3), 1988, p. 209.

[19] Hughes, R. J., On Feynman/s Proof of Maxwell's Equations, Am. J. Phys. Vol 60, No. 4 1990 p 209ff.

electromagnetic; and it supports Einstein's contention that there was a connection. It should then not be necessary to prove that Maxwell's laws are supported by attribute theory. This is important because we would like to assert that our radial function for the atom, after being inverted in unity, is a solution to the curvature of Einstein's field equations in general relativity. This curvature could theoretically be obtained from iteration of N in an equation similar to equation 4-6 if it were inserted for the radius into the Friedman equation. Equation 4-6 is not exact however. In fact, radii predicted by the equation are off by an average of minus 24%. I cannot accept the inverse of the parametric equation which are valid for atomic case but are off by -13 to -48 percentage points. The de Broglie solar equation, $N N_{Av} hV = Gc$ requires knowing Newton's Law to find solar attributes.

Consider a photon emanating from a star. As the photon moves out it globally encompasses a greater mass of stars as it goes. The star has its own central mass; but, the central mass felt by the photon is, in time, continually increased by the mass it now globally encompasses. By Newton's Law there is produced an ever increasing field against which the photon must move, losing energy, as frequency, and picking up red shift as it goes. The photon continues at the speed of light until it reaches earth. At this time a redshift is measured. A redshift, a velocity, a field, and a distance, a density, and possibly some attribute are involved in the process. Surely it moves consistent with the Hubbell constant as defined in Friedman's equation by the square of velocity divided by the square of distance. The density of the global volume encompassed is known only through the Hubble constant. The velocity and density are not measureable. The units of the Hubble constant can be expressed as frequency squared. It would be neat if density were the known proportional to the change in frequency squared!

There are other means of finding distance but the velocity is knowable only if the density and distance are known. Certainly velocity in the Hubble constant is not a velocity of outflow as is claimed. This is true because Friedman's equation, and Einstein's as well are correction to Newton's Law. Newton's Law has no velocity directed outward. That velocity is directed tangentially. Solar systems have attributes or quantum gravities if you will, but, attributes do not lend themselves to characterize inter galactic distance.

Dark energy is currently an object of great study. The Hubble constant was the impetus for the conjecture that the universe is expanding. Warren Arp was an assistant to Hubble who disputed this. For this he was ostracized in the States so greatly that he took refuge in the Max Planck Institute in Germany. The equation that is used to evaluate the Hubble constant is:[20]

$$\frac{1}{a}\frac{d^2a}{dt^2} = -\frac{4\pi G}{3}\sum_i \rho_i(1+\omega_i)$$

4-7

---

[20] Ensley, J., *Elements*, Clarendon Press, Oxford, © 1995, p. 1ff.

The first and all terms in the above equation are really in units of the Hubble constant. All terms have the value of the Hubble constant in the equations of general relativity. Without resort to further equations, the denominator of the Hubble constant is a distance squared, *which distance is the radius of the field* through which the photon has traversed; and, the numerator, contrary to popular belief, is the *velocity on the radius of that field* and definitely not the outward velocity of the emitting star. Warren Arp was correct. The habit that cosmologist have of inventing names and weird units for large numbers obscures our ability, couched in terms of conventional units, to comprehend. They apparently dislike powers of ten notation.

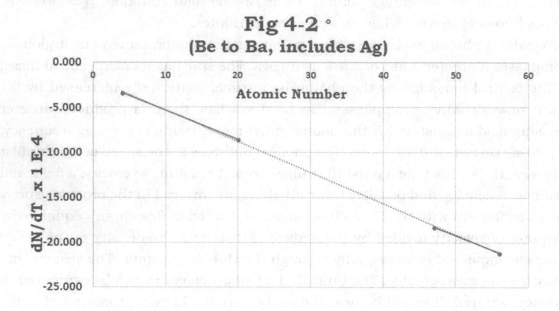

Fig 4-2 °
(Be to Ba, includes Ag)

As in the case of the electrons in the atom, all the planets have elliptical orbits. There is a circular orbit for every elliptical orbit each with the same energy. We were able to define limits of the ellipse from knowledge of this variation in motion that was noted earlier. The times of revolution for some cosmic bodies are in mega-million years; so, their elliptical path is never mathematically relevant.

We are now ready for an easier approach to finding the variation of attribute with temperature. First we calculate the expansion from room temperature to 5000 degrees Fahrenheit. Let A be the linear coefficient alpha.

$$R_{500} = R_{20} + R_{20}A(4920)$$

4-8

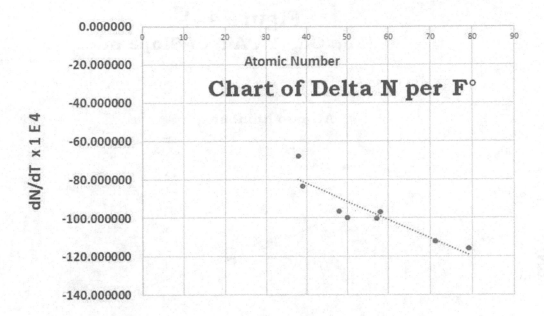

Fig 4-3
From Left (Sr, Y, Cd, 2d: Sn, 4p
La, Ce, Lu Rare E; & Au, 3d.)

Next one simply sets up the aufbau matrix for the element in question and iterates for the value of N for your radius at $5000\ ^0$ F as you would for the radius at $68\ ^0$ F. I then find the slope, m, of the delta N per degree by simply dividing the change in N by the change in temperature. The slopes were about the same for different areas of the table; but, the intercepts were different for different areas of the table. I could not get good results for the p orbitals because they were liquids or gases. One could get volumetric data for the p liquids and divide by three perhaps. The s orbitals gave the sharpest results. The three d orbitals gave a similar slope to the f orbitals with a slightly different intercept.

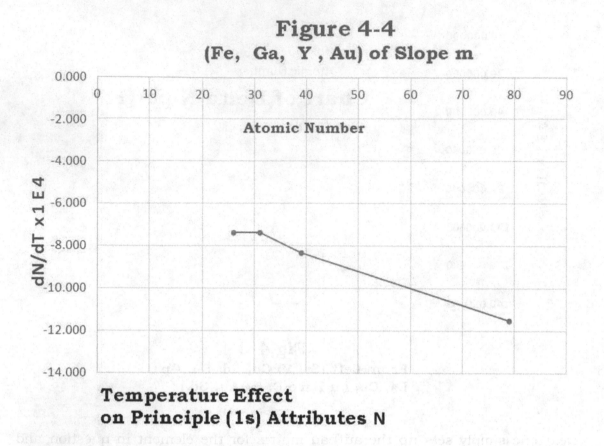

Figure 4-4
(Fe, Ga, Y , Au) of Slope m

Temperature Effect
on Principle (1s) Attributes N

When here is a change from the room temperature value then there is a shift in attribute number, changing the radii, and affecting the density. One not only must shift the set of attributes to satisfy the linear expansion coefficient, α, which is tantamount to shifting the values of N to accommodate the temperature. We are working with linear expansion coefficients. Uranium does not seem to fit into the range.

Chapter Five

The Strong Force

The investigators that described the weak force to be an electromagnetic force, Weinberg and Salam, also indicated that the strong force was probably electromagnetic as well. This chapter is designed to affirm that conjecture. But we can include gravity as well. The next chapter will deal will some aspects of the weak force which are perhaps new, and go into the ramification of the expression of protons, neutrons, electrons, neutrinos, and other particles of the zoo as orbital systems. In my book Quantum Gravity and the Unified Field in 2000 I described orbital systems that would allow such particles to be described in terms of relativistic orbiting masses that could verify their published masses in mev.

The idea at that time was to think of multiple electric charges as being one third the magnitude of the charge and mass of the electron. Thereby we can affix the -1/3 and +2/3 charges of the downquark and upquark respectively as postulated by Murray Gell-Man. Suppose a particle containing three such negative charges is orbited by six positive charges it would have the overall charge of three of these small positive charges or the charge of a proton it would have the UDU configuration of a proton. Take away three orbiting charges and you have the DUD configuration of the neutron. One could not expect a proton to go to an electron if their centers were dissimilar? It has been claimed that there was an electron inside the neutron. There isn't. The electron comes from pair production where the positron degrades to an anti-neutrino by leaving three positive charges with the neutron to be acclimated regarding mass. Actually the neutron is heavier and I will show this using a pair of orbiting mechanisms that are similar to the unified field of the atom. This should be evidence that the attribute orbiting mechanism is applicable to sub atomic phenomena.

Without better knowledge of the relative masses of the six positive charges I am going to assume that they are each 1/6 of 1/1836.15 of an electron mass. Notice that the proton structure will be like that of carbon so we now have a new nuance for descriptive nuclear chemistry. We expect that we should be able to form single, double, and triple bonds. The neutron has the structure of lithium and should be able to participate in the bonding as well. So without a doubt we have established the outline of an electromagnetic nature of the strong force in a little more classical role than might have been anticipated heretofore. I was able to determine values for the three types of bonds mentioned in the year 2000 by a method that is not probably going to be improved upon currently. I will have to refine the method before I can recommend it for atomic bonding. These values are given in Table 5-1 below:

TABLE 5-1

MEV VALUES FOR NUCLEAR BOND TYPES

Bond	Single	Double	Triple
Proton-Proton	2.05098	1.338437	.873442
Neutron-Proton	4.76538	2.505907	1.317748
Neutron-Neutron	2.86730	2.385196	1.98415

Notice that the double bond strength is the geometric mean of the other two bonds. Ostensibly, one should be able employ the bond energies of Table 3.2 to a wider variety of nuclear reactions. There may be a temperature effect that needs to be assessed. The nuclear bond results should be compared to current nuclear shell model calculations. These values agree with the literature.

TABLE 5-2

Nuclear Reaction from Estimated Bond Energies

Nuclear Reaction	Literature Energy	Calculated Energy	Percent
$H_1^2 + H_1^2 \rightarrow H_1^3 + H_1^1$	4.03	3.88	-3.6
$H_1^2 + H_1^2 \rightarrow He_2^3 + n_0^1$	3.27	3.24	.8
$H_1^2 + H_1^2 \rightarrow He_2^4$	17.59	16.93	-3.5

Six nuclear reactions were tested to see if a linear combination of the above bond energies would satisfy the energy requirements. If energy is given to the orbital system of protons or neutrons one can find particles of the zoo. This is best covered in a separate chapter which will deal with that subject. The weak force ostensibly elicits an electron from the lighter neutron. The neutron then becomes a heavier proton. It has been presumed that the electron came from the proton. In fact it comes from a pair production wherein the other half of the pair, the positron loses three small charges and becomes the other product, an anti-neutrino. We will next, however, show an interesting calculation that describes the situation in the proton and neutron whereby the neutron is heavier than the proton; however, it does not house the electron.

The model assumes that the proton and neutron have radii of the order of 10^{-15} meters. To bring this about it is assumed that the equation below applies to both:

$$r = 5.05x10^{-8}\frac{\alpha \ln(\alpha)}{\sqrt{N+6}}$$

5-1

The essential mass data for the proton and neutron is tabulated below. The number of orbiters and their masses is pertinent because we will judge the difference in the relativistic masses of the orbiters to be indicative of the difference in mass of the proton and neutron. We also assume that the usual Poisson quadratic will be evaluated at N=6 and N=3 of the proton and neutron N values that predict the correct relativistic factors used to determine the difference in masses of the two nucleons.

Table 5-3
Masses in Kg

M proton	1.6726217770E-27
M neutron	1.6749273510E-27
M neutron-M proton	**2.3055740000E-30**
M electron	9.1093897000E-31
Me 6 orbiters 2 el'trons	4.9611359094E-34

Below in Table 5-4 are the data from which we can arrive at the difference of the masses of the proton and neutron. In calculating Table 5-4 some differences were used from normal atomic calculations. The central mass (G) of the proton is the M proton of Table 5-3 minus 6 orbiters; and that (G) of the neutron is M neutron minus the mass of 3 orbiters. The central charge is the charge on one electron that is three thirds of an electron in both cases. The actual Poisson (not shown) is calculated as usual, but with only a sixth and a third of the N values shown. The square of radius is based on equation 5.1 and the square of velocity is based on the amended G's of (GM+KQ)/R. This permits calculation of a relativistic factor for mass of orbiters that is iterated until they sum to the required value of the difference of the masses of the proton and neutron. The proton was accelerated until the relativistic increase was only 2 in the sixth place. The average of the two radii at the point when the relativistic difference in the masses of the orbiters was equal to the difference in the masses of the two nucleons was $7.061x10^{-15}$. Note that the positive orbiters of the proton have the mass of two electrons.

Table 5-4

N	Alpha	Radius	V²	Velocity	R Factor	R Mass
715160	9999902	4.39e-16	1.355e16	1.6405e8	1.1775	2.143e30
715900	999992	4.38e-16	1.79 e16	1.134 e8	1.167	2.305e30
890000	9999918	3.31e-16	7.148e16	2.6755e8	4.886	4.51 e30

One can question whether this average radius and the two elements averaged are indeed different and indicative of the actual case. The value of radius of the proton occurs in two levels five N apart. The multiplier of the radial functions of the proton and neutron instead of being -10 was minus -3.79535734 x 10^{-8} which was required to obtain .8775 fem., the radius of proton found by the Max Planck Institute and others recently. It served as well for finding the radius .33 fm. of the neutron. The excellent agreement in radii of the two nucleons is more than hoped for. The Poisson (not shown), which acts on the velocity squared, was calculated using the normal quadratic. I originally thought that I could use one third N for neutrons and two thirds N for protons. The value required for the fractional effect of the charge on the orbiters was .34682145 for N for the neutron and half of that for the first orbit of the proton. One would expect that there would be different values of N because of the lesser charge of the orbiters. The central charge was that on an electron. The Poisson was calculated on the quadratic in N, but with the above fraction (of the value of N which is usually used fully for atoms. The radial difference of the proton and neutron is quoted as .33± ca .17 fm.[21] I find essentially the value of .33 fem. in the estimate of Table 5-4 which is the upper value quoted in the Phys. Letter reference. Experiments were carried out at the Paul Scherrer Institut (PSI) (Villigen, Switzerland) which is the only research institute in the world providing the necessary amount of muons. The international collaboration included the Max Planck Institute of Quantum Optics (MPQ) in Garching near Munich, the Swiss Federal Institute of Technology ETH Zurich, and the University of Fribourg, the Institut für Strahlwerkzeuge (IFSW) of the Universität Stuttgart, and Dausinger & Giesen GmbH, Stuttgart. The value they obtained was .8775 x 10^{15} meters.

The foregoing is taken as evidence that the quarks involved in proton and neutron confinement constitutes an electrochemical series of events. This should dispel any doubt as to the electromagnetic nature of the quarks. When it comes to the muon combinations of quarks and anti-quarks we will demonstrate in chapter six the nature of enhanced relativistic orbiting masse smaller that electrons or positrons will help divulge the nature of the particle zoo. Yes mass is still a factor exhibiting gravitational effects.

---

[21] *Phys. Rev. Lett.* 108 1-2507, (2012).

Maxwell demonstrated the unification of the electric and magnetic fields. Including the definitions of the strong and weak fields, there are now four accepted forces in nature. Gravity has eluded attempts to unify with electromagnetic fields. In 1918 Hermann Weyl made an unsuccessful attempt in which he was first to coin the term "gauge invariance".[22] C.N. Yang and R.L. Mills derived a theory from a local symmetry requirement in 1954.[23] Steven Weinberg stated in 1975 that no one had been able to include gravitation in the family of particles of a generalized gauge group that mediates the electromagnetic, the weak, and perhaps also the strong force.[24] More recently Richard Feynman's derivation of Maxwell's equation from the assumption of a quantum gravity has been both criticized [25] and praised.[26] The paper was written in 1948 but never published. Unified field theory deals with charged masses in orbit. We are going to consider the orbital system of quarks in the nucleus in this chapter.

From 1988 to 1983 I authored three papers dealing with a viable unified field.[27] [28] [29] Speculation as to the existence or nature of quantum gravity and the lack of detection of the postulated graviton has led to many attempts to unify three of the four supposed forces of nature into a grand unification, or GUT theory. Weinberg, Salam, and Glashow unified the weak and electric forces under the electroweak theory.[30] [31] Weinberg states that the photon is the most visible member of the family of elementary particles required by a generalized gauge group that mediates the electromagnetic, weak, and strong interactions. From definitions of the electroweak force, which merges with the strong force at higher energies, it should follow that the strong force is also electric at lower energies. The strong force field describes the force holding the nucleus together. The rate of decay of the hadronic particles characterizes, in part, the difference between the strong and the weak field.[32] The common practice of using probabilities in quantum mechanics befits such a kinetic

[22] Weyl, H., Theory of Groups and Quantum Mechanics, Dover © 1930, p. 100.

[23] R. Feynman, The quantitative behavior of Yang–Mills theory in 2+1 dimensions, Nuclear. Phys. B188 (1981), 479–512.

[24] Weinberg, Steven, *Physics Today*, **28** (6), p. 37 (1975)

[25] Dyson, F.J., Feynman's Proof of the Maxwell Equations, *Am. J. Phys.*, 58 (3), p. 209 (1990).

[26] Hughes, R.J. On Feynman's Proof of the Maxwell Equations. Am. J. *Phys.* **60** (4), p. 301, (1992).

[27] Brooks, J.O., Electrogravimetric Field of the Classical Atom, © 1988

[28] Brooks, J.O., Newton Coulomb Field of the Atom, © 1991

[29] Brooks, J.O., Atomic Attribute Theory, © 1993.

[30] Weinberg, Steven, Light as a Fundamental Particle, *Physics Today*, **28** (6), p. 32 (1975)

[31] Salam, Abdus, Gauge Unification of Fundamental Forces, *Science*, **210**, (11), p. 273, (1980).

[32] Guillemin, Victor, Story of Quantum Mechanics, Scribners, New York ©1968, p. 151

description. This has resulted in new types of quantum numbers such as are used to describe quarks. Conservation of strangeness, for example, only determines what may not happen, not what does happen. Particles created via the strong force decay 10^{13} times more rapidly by the strong force than the 10^{-8} sec decay time of the weak force. The weak force is the one that preserves strangeness via this relatively slow decay time. Many of these conservation laws are in force because of Noether's theorem. We have dispensed with that theorem because it was spawned for a Lagrangian system which her premise is destroyed by Newtonian systems. We shall deal with a semi-chaotic nucleus that is not necessarily a portion of a beam.

The following tables reflect the calculation of field quantities for strong field orbiting quark systems of the proton and neutron. Think in terms of integral thirds of charge of the electron as an electrino (or positrino). Consider the central charge on an upquark to be an electrino which will be orbited by three positrinos. A downquark consists of only one electrino. By the new proton configuration six positrinos orbit three electrinos. The neutron configuration has only three positrinos orbiting three electrinos. A downquark consists of a single electrino while an upquark is a down quark orbited by three positrinos. Since fractional charges seem real, there must be an associated mass entity. The field per positrino is of the order of 10^{40}. The calculations except for boundary values mimic those we can set up field and frequency calculations from the N-matrices similar to those described for the atomic calculations by equation 1.6. The nuclear N value changes from one shell to the next shell by the methods used in Chapter 3. The N-value level of spacing between shells and sub-shells is simply not as great in sub-atomic systems as it was in the atomic case. Tables 3.1 and 3.2 that follow depict this. The attribute levels correspond to the same aufbau pattern as was found for the atom. However, there attributes have a slightly different shell differential.

Table 5-5 The Proton

N x 10^4		10^{39} x g_{mass}	10^{40} x g_{charge}	Velocity m/sec
$N_{1,1}$	479998	8.44	1.51	299538889
$N_{1,2}$	480008	8.44	1.51	299552254
$N_{2,2}$	480010	8.44	1.51	299554928

The attributes of the neutron are similarly described.

Table 5-6 The Neutron

N x 10^4	10^{39} x g_{mass}	10^{40} x g_{charge}	Velocity m/sec
$N_{1,1}$ 479245	8.385	1.4925	299538889
$N_{1,2}$ 479295	8.389	1.4937	299552254

The test particle is the positrino, which is assumed to have the charge to mass ratio of the positron. The radius of the first orbiting positrino in both the proton and neutron is of the order of 3.8 x 10^{24} meters which changes the boundary value of equation 1.2. The calculation that characterized gravitational and electrical fields inside neutrons and protons depends on a model with a negative nuclear triad of electrinos surrounded by three positrinos for the neutron and six positrinos for the proton. In recombination studies the field, E, has been a variable during collection. The neutron is similar to lithium in structure and the proton is similar to carbon. Nuclear bonding can be structured similar to chemical bonding.

The nuclear bonding structure mimics elements three and six. The latter, a proton, is an ionic complex. It is natural to assume this is the basis for nuclear reactions as there will be three categories of bonds each category having single, double, and triple type bonds. The quantum mechanical nuclear shell model assumes that the nucleon fills nuclear energy levels similar to the way electrons fill atomic energy levels.[33]

The method for detailing nuclear bonding is modest. Perhaps a procedure to transmute a base element to a scarce element by selectively severing bonds can be devised. Using the present procedure for bond strengths detailed in chapter one it is possible to estimate the energy of a half dozen nuclear reactions to the order of 3 %. Equation 5-2 gives the parametric equations for strong force

Using the definitions of an α of equation 1.2 we must find an N value near the $N_{i,j}$ of Tables 5.1 and 5.2 that give the correct mix of the nine possible bond types and satisfy the nuclear shifts and energies of the nuclear reactions. The spreadsheet for doing this is complex, nevertheless, one can iterate the conforming mix.

[33] Johnson, Karen E. Maria Goeppert Mayer, Atoms, Molecules and Nuclear Shells, *Physics Today*, **39**, (8) (1986).

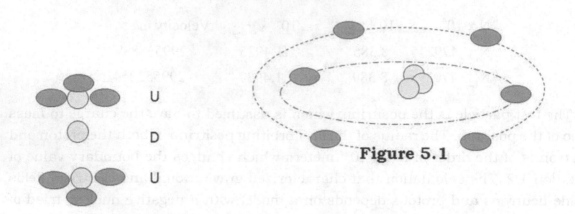

Figure 5.1

The Proton as Described by the New Quarks
Center 3 x (– 1/3), in orbit: 6 x (+ (1/3)

Recall that masses and charges are only a third of these values for an atom. The relativistic correction is applied just once; else, one has positive feedback. The attribute radial function of equation 1.2 becomes:

$$\nabla^2(\phi) = 9N^2 - 271N + 588$$

5-2

The attribute unified field function of equation becomes:

$$g(N) = \frac{(GM + KQ)\nabla^2\phi(N)}{r^2(N)}$$

5-3

Calculation of frequency is as in equation 5.4.

$$f_{i,j} = \frac{\gamma \pi^2 m_e v_e(N)c}{r^2(N)\sqrt{1/1 - v^2/c^2}}$$

5-4

This means, among other things, that the relativistic mass increase does not participate in gravitation within the orbiting frame.

The fit of the above equations is used in calculating the relativistic change in hadronic mass of the particles from which neutrons and protons decay is given in the following Table 5-7. There is a converging set even though the attribute number difference in N is very small. This results in certain arbitrary assignments because the energy levels of the positrinos are not broad. Enough of the particles heavier than nucleons pick up relativistic energy are surmised to be masses, either protons or neutrons, imbued with relativistic energy. Masses quoted in table 5.1 are relativistic for the orbiting particles. Protons or neutrons become heavier hadrons as the attribute, N, increases.

TABLE 5.7

Production of Hadronic Mass via Increase of Attribute

Attribute Proton N x 10⁴	Attribute Neutron N x 10⁴	Mass Proton Mev	Proton Percent	Mass Neutron Mev	Hadron Neutron Percent	Mass Particle Mev
490052	480113	1116.2	.0	1115.9	.0	Λ⁰ 1115.6
490066	480131	1181.8	.6	1182.4	.6	Σ⁺ 1189.4
490078	480122	1192.5	.0	1191.6	.0	Σ⁰ 1192.5
490080	480123	1202.7	.5	12.01.0	31	Σ⁻ 1197.3
490088	480133	1313.1	.14	1309.1	.45	Ξ⁰ 1315.0
490090	480134	1327.3	.48	1321.6	.04	Ξ⁻ 1321.0
490124	480154	1673.9	.11	1684	.72	Ω⁰ 1672

This involves four by four matrices. There is the order of five one thousandths of an attribute difference between rows for the neutron, and twice that for the proton. This is over and above the attribute to yield the quoted Mev values for proton and neutron plus the increase to achieve the requisite mass. One doubles the hadronic frequencies after they are prorated for occupancy. The "as is" frequencies are taken as the bond strengths when the match is done. The range of attribute, N, is within unit spin difference of the principal attribute. Spreadsheets can be devised to iterate this, changing only the value of the principal attribute.

The next table is the result of iterating correct bond strengths in Mev for the nine cases of nuclear bonds that will also agree with the reaction energies in three nuclear equations. The blanking procedure of chapter one is applied using equation 5.4. This involved finding a linear combination of nuclear bond breaks and bond makes for the reactions in question. Other types of orbiting systems involving quarks can be utilized because an iterated N is available to match virtually any energy by adjusting the attribute. This is essentially an existence hypothesis.

The phenomenon of symmetry breaking is one of the scale of the reaction. One could think of any r-values as a string and the concatenation of frequency as embellishing the energy using several strings. The current practice in strings is to work with ten dimensions. This should consist of sufficient scales to cover cosmological, astrophysical, solar, earth scale, atomic, nuclear, electronic, mesonic, and a pair of sub-mesonic systems.

In atomic bond estimation one must do matrix multiplication of the anti-logarithmic, base ten frequencies to assess bond strength. To do so on a nuclear scale requires a leveling factor to the order between one and ten .The numerical attribute difference is a thousand times smaller than it is in the atom. As orbital radii decrease the role of gravity diminishes. The leveling factor for neutrons and protons is given below.

$$R(N) = -10^{27} f(N)$$

5-5 b

Either Newton's universal constant or Coulomb's law constant can correlate the force of gravity at the earth scale.[34] Earth gravity as a corollary of the atomic field of the atom is thus a function of Coulomb's law and one other constant.[35] We should remember that a natural law is what nature does, not what compels it. Field theory is primarily a theory of dynamics and not kinetics. We may describe elemental frequency matrices or their nuclear counterparts in terms of the associated unified field matrices of the radial shell matrices.[36] Matrix manipulation consists of interaction of the frequency anti-logarithms by multiplication of anti-log entities. The matrix frequencies are multiples of the leveling factors of frequencies and convert the interactions to energy units.

Unification at decreased orbiting radii still exists. This means that although extremely small particles are driven by electromagnetic energy there must be a slight gravitational contribution as long as a charge to mass ratio in effect. It is dynamically odd that the weak force particles have such massively disparate intermediate energies. The W± and Z particles mediating the weak force have greater energy than the mesons (now termed gluons) that mediate the strong force. Some of the nomenclature designed to preserve symmetry conveys concepts are antithetical to premises of attribute theory. There is then the possibility that one unified field does describe all of nature's fields. Energy above the rest mass does not create mass that gravitates. This is important to recognize in the case of the macro-cosmological model where relativistic velocities are really small except for their chemical entities, which are still at atomic scale.

The quark is a well-defined fragment. Up and down are acceptable prefixes for quarks. Strangeness, or charm, seems to imply a two quark, positrino poor orbital system. Truth and beauty may be positrino-electrino enhanced strangeness. We have shown at least a procedure successful in the atom to be a viable means of energy storage based on relativistic orbiting entities. The energies involved mimic the energies of named particles. Indeed the method could attain the energy of any observed particle. Within its scale mass is tied to its test particle velocities. Charge is probably a function of mass until the macroscopic matter is met with.

There is then a good possibility that one type of equation describes all of nature's fields. All those who set forth theories have experienced difficulty in extending their findings. There is usually some shortfall, some criterion not supported by physical fact or some widely accepted postulate that does not conform to the extension of the

---

[34] Brooks, J.O. "Validating Mass in the Unified Field", AAPT, Indiana State University, 1992.

[35] Brooks J.O., Quantum Atomic Gravity in Relation to Solar Gravity, © 1991.

[36] Brooks, J.O., Newton Coulomb Field of the Atom, © 1991.

theory. Newton had difficulty in extending gravity to light, and yes, his prediction of the precession of mercury was off by a factor of two.

Einstein did not contribute to quantum theory after 1925. His effort was to extend special relativity to general relativity. He used a d'Alembertian in space-time within a Riemannian geometric.[37] This required using a tensor with ten elements[38] in this metric he concluded in 1932 that $g_{u,v}$ and ϕ_u, the gravimetric and electromagnetic potential variables, do not lend themselves to a continuum. I found these to have the same Poisson. However, in 1912, he had shown that electromagnetism replaces gravitation within a spherically bounded three-dimensional space having a cosmological constant of $\lambda = r/4$.[39] Einstein concentrated his effort on the cosmic, or macroscopic, scale so that neither the simplifications of a smaller source nor much of its data were available to him. We know the state of mathematical calculations in the fifties does not compare with advances inherent in even the PC spreadsheets of today. Time for requisite multitudinous iterations was not on Einstein's side. Einstein's Equivalence Principle requires the extension of special relativity into a corresponding acceleration, in a reference field of small curvature. Thus special relativity becomes general relativity. General relativity is not as yet fully accepted beyond the theory stage. In a later chapter on the Hubble constant we shall employ Einstein's notation for general relativity in concert with a quantum gravity to arrive at a neat expression that supports general relativity. In this respect, it may be considered a new solution to Einstein's equation incorporating quantum gravity and involving attribute theory. Certainly one could not hope for a simpler macroscopic expression of quantum gravity than an equation relating Newton's universal gravitational constant to the quantum variables of the Rydberg equation. The addition of energy bringing out the test particle into a tighter orbit really would appear to break any symmetry of kinetic and potential energy expressed in the Hamiltonian. In the atom and in the nucleus higher orbit is higher energy orbit but at a lowered radial level according to equation 1.2. In the atom, the energy input shortened the potential but increased the kinetic energy because of the heightened frequency.

There is a neat proof of the equivalency of gravity and electromagnetic theory. The equivalence of gravitational and coulombic force of chapter can be greatly simplified if one denotes the ratio of the charge to mass on the sun as Q_{sun}/M_{sun} is a constant and note that there is another constant such that q_{planet}^-/m_{planet}. Then one can write that what is to be proven:

---

[37] Pais, Abraham, Science and Life of Albert Einstein, Oxford Press, Oxford, © 1982, p. 349.

[38] *ibid*, p. 273.

[39] *ibid*, p. 287.

The attribute radial function of equation 1.2 becomes:

$$\frac{GM_{sun}m_{planet}}{r^2} = \left(\frac{Q_{sun}}{M_{sun}}\right)\left(\frac{q_{planet}}{m_{planet}}\right)\frac{KQ_{sun}q_{planet}}{r^2}\left(\frac{M_{sun}}{Q_{sun}}\right)\left(\frac{m_{planet}}{q_{planet}}\right)$$

5-6

Since it can be shown that:

$$G = \left(\frac{Q_{sun}}{M_{sun}}\right)\left(\frac{q_{planet}}{m_{planet}}\right)K$$

5-7

Therefore Newton's law equals itself, and since the end pairs of the RHS of equation 4-6 cancel we can assert that:

$$\frac{GM_{sun}m_{planet}}{r^2} = \frac{KQ_{sun}q_{planet}}{r^2}$$

5-8

Which is what we wanted to prove. Because the product of Coulomb' law constant and reciprocal canceling constants is equal to Newton's universal gravitational constant then equation 5-6 is a two way identity and equation 5-8 is one of them. Thus Einstein's value of the *cosmic constant* within a sphere where the equality of gravitational and electromagnetic fields is unity has the value of λ/4. So, taken with the Rydberg Rearrangement then we have two independent sources that assert the equality of Newton's universal law of gravitation with Coulomb's law in the solar system. In the process Einstein's λ/4 is equivalent to the Poisson in our solar system. The value of λ in our solar system is 4. Could this mean that our solar system is 4 dimensional? The utility of the Poisson used with atomic attributes is probably not going to be extended to solar realms because Einstein's cosmic factor will probably depend on local conditions. Investigators may wish to examine brane theory, Kaluza-Kline, or Yang-Mills theories to try and make sense of the constant transforming G to K of chapter 1one. I should think that two galactic entities having different charge to mass ratios each would move independently, as our Sun does, in possibly with up to seven different motions. Fitting two galaxies into a format that would satisfy general relativity does not seem likely to me. General relativity, it is said, need to know where all the masses are. Nothing is ever said about how big a pattern of sky constitutes the domain of general relativity.

The Variational Principle in Riemannian geometry requires a four-vector. In the cylindrical coordinates that we have been describing only a two-vector is needed. Since attribute theory is equivalent to Schrödinger's equation in cylindrical coordinates there is a real advantage in its use. It greatly simplifies the world line. This is also the reason that both scalar and vector components have the same form in cylindrical Bessel functions.[40] Einstein's 1912 effort implies that, given

---

[40] Reitz, John R. and Frederick J. Milford, Foundations of Electromagnetic Theory, 2nd Edition, Addison-Wesley, Reading, Mass., © 1968, p. 109.

a common source, in a spherically bounded Poincare universe, the gravitational and electromagnetic vector system is equivalent and satisfies the equivalence and variational principles.[41] We may conclude therefore that the atom and the nucleus have the properties similar to this type of universe. Compared to the atomic the sub-nuclear frames they have an apparent difference in their aufbau, or in the way relativistic energy accrues to the system. We apply the Equivalence Principle to assess the attribute boundary frequencies of the rest point masses of the nucleons.

In the atom we iterate the numerical attribute for known radii. In either case we arrive at boundary attributes for that system. The rationale that equates gravitational energy with non-inertial mass includes spin which is a temperature dependent variable that oscillates ever more greatly between two energy extremes whose attributes become less at higher temperature. This is discussed more fully in the chapter when spin is discussed.

How else would you interpret the time average de Broglie frequency as that given by the geometric mean of velocity or orbit and speed of light? The photon energy mediates the increase of energy as frequency in atomic systems. In the nuclear case the increase in energy is again due to a frequency increase as the electrino goes relativistic. Due to quantum jumps, the geometry is affine and the matrices that describe it in attribute theory are at least pseudo tensors.

The phenomenon of spin is analogous to the Einstein-Cartan coefficients linking torsion to spin.[42] Their torsion spin tensor requires eighty elements. One could also report simple spin at a temperature by means of elliptic integrals in real time.

Einstein steadily maintained that quantum theory was not complete. In his debate with Bohr he insisted that a complete quantum theory would not need to include a classical apparatus limit which Bohr had suggested was needed to observe the quantum object directly.[43] Quantum theory would predict such an observable from an earlier observation of material interactions. Gauge symmetry breaks spontaneously when a physical state becomes unrealizable.

The present state of *attribute mechanics* has a link to a portion of the Schrödinger equation and has a steady state. A jump in time occasioned by the entry of spectral energy provides a history in time. Its symmetry holds over a range of smaller size. The inversion of the parametric equations extends to solar size the radial function and the Rydberg Rearrangement does the same for the field; so, I would argue that we have symmetry, and classical connections from atomic to solar scale. Cosmic scale is a combination of solar scales. Apparently the Lagrangian is classically needed to describe a combination of solar scales. Since Einstein specifies a cosmic

[41] Einstein, Albert, *Nature*, **8**, p. 1010, (1920).

[42] Hedhl, F.P., G.D. Kerick, et al, *Rev. Mod. Phys.* **48**, p. 393 (1976).

[43] Widom, Alan, Letters, Physics Today **47** (1), p. 67 (1994).

constant in order to embrace the equivalence of electromagnetic theory and gravity this may be just the Poisson.

Accounting for strangeness will probably involve time for the initiation of a mesonic 2-quark orbiting system from a degenerating hadronic 3-quark system. This is like the kinetics of two colliding entities that involves rate determining steps. Another anomaly is the concept of quark confinement. Surely the concept of a state function rather than a path function will reduce some of the difficulties to more manageable proportions.

Perhaps it would not be oversimplification to circumvent the SU (2) ⊗ U (1) process unless you were attempting to predict how a specific ensemble would react when placed in reasonable proximity in real time. Any entity in the ensemble will react if in reasonable proximity in time. If an entity were not of suitable scale it could result in an infinity that renormalization might or might not cure.

The approach of radial attribute theory is that energy delivered will result in energized states that are either stable or will decay. The as yet undetected entity with one-third the charge of an electron is still believable. In any orbiting system such as the quark, which has a nucleus with orbiting positrinos, the particulate system cannot be fundamental. The fractionally charged quark is bona fide. Since Millikan used an ordinary electron he could not be expected to find fractional charge. Note that Millikan's method also involved the charge in an electrical and a gravitational field on a particle with a definite mass and charge. If one takes the electron to be the smaller anti-particle of the proton it may be divided three particles each with two negatives orbiting a positive. The anti-particle thus has only a reduced central mass.

We assume that the mass of a quark is one third of the proton mass in electron mass units. In the neutron there are three positrinos in orbit about an electron or three electrinos. The particle is neutral. The upquark results from the acquisition of three positrons making it into an upquark and giving it the charge positive 2/3 q. It thus has six positrinos in orbit and mimics the chemistry of carbon. This is the maximum number of particles found by J.J Thomson to be stable in orbit.[44]

It is equally important that the neutron, having three electrinos or one electron, orbited by three positrinos, has a lithium-like outer structure. Bonding between nucleons furnishes us with a reason for nuclear stability. It seems likely that, whatever the colliding beams in super-colliders are, the process either starts from or decays to a proton or to a neutron. The practice of using Mev to report masses is convenient. It is also misleading if one tries to use the mass equivalent of the particle's energy in self-gravitation. If Einstein's equation for mass increase with velocity were employed several times it would violate the second law and the positive feedback would send the mass to infinity. Whatever energy or relativistic mass particle detectors measure reflects an energy buildup and subsequent decay. The

---

[44] Heilbron, John L., and J.J. Thomson, Bohr Atom, *Physics Today*, **24** (4), p. 23 (1971).

common baryons up to 1672 Mev become excited states of the proton or neutron. The symmetry-breaking pattern is the radius and so the radii predicting attributes offer a measure of symmetry.

Is the $W^{\pm}$ or the Z^0 from a neutron or a proton? The mesonic or two-quark system can have two modes of orbiting. One mode could be about the center of mass of the pair. Another mode would be comprised of DD, UD, and UU quarks of charges minus one third, plus two thirds and plus four thirds. Each pair would have two electrinos in the nucleus and zero, three, or six positrinos in orbit. Many questions remain unanswered.

I have in this chapter tried to present a means of replication of known facts regarding the protons and neutrons using the extension of the methods used for the atoms that are inspired by the parametric equations. Facets other than those covered here may be found in other chapters. I am following the criterion that says extending one's method to as many applications where they appear to be successful will tend to not only validate but render the method useful for further extensions.

Chapter Six

Weak Force Attributes

The attribute parameters constitute a real radial solution to Schrödinger's equation. This chapter will augment the earlier exposition of the calculated mass of the proton and neutron. We are really just redefining Feynman diagrams with a slight twist. Symmetries can be geometric or dynamic. Other Each entity has a unique set of attribute numbers that characterize their shell and subshell orbitals. Symmetries have internal symmetries in the multiple dimensions of the Poincaré and Lorentz groups.[45] Great simplification ensues when using attributes. Quantum mechanics is an ad hoc classical theory that relies on solving the wave equation only in probability terms. There is no doubt that Quantum Mechanics has had success and will continue to do so especially in kinetic areas because there is a probabilistic approach to chemical reactivity. Some procedures result because they have bolstered theory in other areas and will be presumed to do so again.

So far we have seen that gravity is apparently an electromagnetic phenomenon. Given this, one method of promoting unification might be to reconcile general relativity with the non-integral quantum gravity aspect of chapter one. Quantum gravity if it exists is non-integral. Wherever Einstein used Newton's universal constant it may be conditioned by use of the Poisson. There currently are no attributes for use in general relativity; but, if attributes can become viable we would have a kind of unification. Relativity and gravity are still going to be incompatible with the quantum mechanics of the Lagrangian and Hamiltonian.[46] Controversy over quantization of gravity remains.[47] [48]

---

[45] Gross, David J, Symmetry in Physics: The Wigner Legacy, *Physics Today*, (48-12), Dec 1995, p. 47.

[46] Jammer, Max, Conceptual Development of Quantum Mechanics, McGraw-Hill, New York, © 1966, p. 133.

[47] Dyson, Freeman J., Feynman's Proof of the Maxwell Equations, *Am. J. Phys.*, 58(3), March 1990, p. 209.

[48] Hughes, Richard J., On Feynman's Proof of Maxwell Equations, *Am. J. Phys.*, 60(4), Apr. 1992 , p. 301.

Heisenberg has noted that in all intellectual progress there is a revision of some concept or previously held position. For instance, one may define the four-vector as a radius with a mass dependent frequency in real space (Lorentz symmetry) rather than a rotator in abstract n-space as Dirac did. A recent article by Haisch speculates that mater may not only be equivalent to energy, matter may be just energy.[49] He describes a ZPF or zero point field that obviates invoking the usual assumptions of quantum mechanics. Andrei D. Sakharov in 1968 proposed the idea that the ZPF leads to Einstein's equivalence of gravitational and inertial mass. He suggested that somehow ZPF generates gravity, but; he never developed the idea further.

This idea of equivalence is inherent in our development of attributes assuming the macroscopic van der Waals criterion which was crucial in demonstrating the real numeric of the equivalence of gravity and electromagnetic theory in chapter one. It is of interest to note that the non-relativistic chaos of the solar system is mild and permits pseudo stable spin variation with almost any orbit.[50] Velocities of the order of the speed of sound are common in atoms. I confirmed that the velocity of electrons in metal atoms are of the order of sound upon adapting cosmological equations.[51] The orbiting particles of the weak force are relativistic just as those of the strong force are.[52]

According to Baggett, the unquestioned acceptance of the Copenhagen version of quantum mechanics has held back progress in the development of alternative theories.[53] The strong force model of positrinos in quark orbit will serve as a model for the electron since the electron will be assumed to be electrino-inos in orbit about a positron. There is a similarity of masses and charge both positive and negative to a lower and lower level. The quote, "the flea has fleas smaller still that bite'em, et cetera ad infinitum" is apropos!

Regarding the anti-particle such as it would be when we just change the sign of the central and orbiting particles and think of the describing not as three UDUP6 quarks but as a UDUE6 quarks. An electron is an udue6 quark-inos with an orbiting central mass which is the ratio of a proton mass to an electron mass lighter than the proton. The udu's are lighter by a factor of 1816 than the UDU's. The use of (–ino) also reflects a smaller version. A small range of attributes describes the range of masses from the field. The contribution of gravity is now *very* small; nevertheless,

---

[49] Herbert, Nick, Quantum Reality, Doubleday, New York, © 1995, p. 42.

[50] Brooks, J.O., Quantum Gravity: Its Radial Attributes, © 1995, AAPT, University of Indianapolis 1995

[51] Brooks, J.O., Validation of the United KG Atomic Velocity Field, © 1993, AAPT, Hanover. 1993.

[52] Brooks, J.O., Role of Gravity in the Strong Force, © 1994, RHIT, 1994.

[53] Baggott, Jim, Meaning of Quantum Theory, Oxford University Press. New York, © 1992, p. 211.

we can claim its unification with electromagnetic theory. As before we employ the Newton in electron mass units and hybridize Coulomb's law to a per electron mass basis. Spin is essentially non-integral. We employ the time average solution and the mass in mev will be calculated from the frequencies of the orbiting-inos. The higher masses will of course have higher frequencies for the particles in orbit that with the proper attribute number, N, we match the mev of any known sub-nuclear entity. It is perhaps enigmatic that cannot foresee or postulate a particle decay that would exhibit reasons for resonances. The resonances at greater velocities in orbit may be an artifact of storing energy in exotic particles.

Positrinos in Quark Orbit

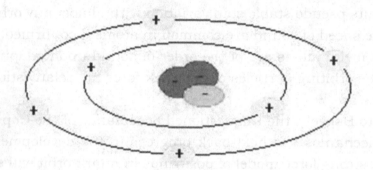

The orbits are a time average equilibrium comprising the radially isotropic sets of field radii. The radial distances between central mass and unit test mass energies define the atomic fields. Figure 6-1 depicts positrinos in quark orbit. It could, with reversal of central and orbiting charges, be an electron entitled electrino-inos in quarkino orbit. As it stands it might also be called positrino-inos in quarkino-ino orbit. This gives further insight into sub-atomic orbital. These matrices of implied curvature is are abstraction not unlike those of the theses that Einstein promoted in general relativity, but with far fewer elements. We achieve the spirit of Einstein's equivalence principal within the framework of special relativity. The price of the unity between atoms is that the first (principal) attributes increase by a sometimes fractional, mostly non-integral amount. The departure of the attributes from their time average gives the atom a spin that is temperature related. Even Mosely's atomic numbers strayed somewhat from being integral.[54]

The unit test particle is the kilogram in solar terms or the unit electron mass in terms of the atom. Since the electrino has an assumed central mass equal to the

---

[54] Haigh, C.W., Moseley's Work on X-Rays and Atomic Number, *J. Chem. Ed.*, 72 (11), 1995, p. 1012.

mass of an electron, the mass of the orbiting electrino could be that of 1/ 1836.15 of that of the electron, there doesn't seem a reason to change those values used in our basic equation. If we base the unified field on the electron-mass Newton, the value still has units of meter/second. There will be a principal attribute for the weak force particles.

We assume that there is Lorentz symmetry. Vincent Icke of Leiden has written a summary of symmetry operations. He details an SU (N) symmetry convention of the weak force in depth.[55] The mechanism is local SU (2) symmetry due to the fermion multiplets or the doublets (e, v_e), (μ^-, v_μ), and (U, D). We consider a nuclear and sub nuclear or quark chemistry that *obviates* the necessity of an SU (3) color charge of the UDU and DUD triplets.

The $W^\pm$ and the W^0 are said to be gauge bosons with symmetry U (1) $\otimes$ SU (2). Our effort tends to question not only the role of color, but also the roles of charm, strangeness, bottom, and top. The resonances of the quark complexes would now seem to depend on the semi-stable relativistic mass jumps in orbit B.

Icke has six reasons why unification is not hopeless. First, all particles should feel the force of gravity and electromagnetism. This *overlaps* whatever the characters of the weak and color forces entail. Second, quarks have sub-multiples, q/3 of charge and there should be *an adaptation of attributes* that would explain this. Third, coupling constants increase in number with the increasing dimension N, of SU (N). Therefore at high masses all forces should merge into a grand symmetry. Fourth, the existences of multiple doublet generations of leptons and quarks that increase in mass indicate that there is more to symmetry than meets the eye. Fifth, exclusion of gravity is unacceptable. It indicates that we require a higher type of symmetry. Since Lorentz symmetry is a peculiar type of rotation of space time, the U (1), SU (3), or SU (3) rotations in some local internal space may involve gravity and spin ½ particle generation. Sixth, there may be an underlying unit between relativity and the quantization mesh. This may involve an elucidation of the role of the fine structure constant according to Icke.[56] He also says that in any quantization of gravity of the solar system Newton's universal gravitational constant may be a function of h, q, and the velocity of orbit which is equal, in MKS units, to v q/cα. This speaks to the first, fifth, and sixth reasons on Icke's list. I do not think that quantization of gravity will be a viable topic even if the real numbered attributes can be figured. There is no energy source that would be available or even useful if it were. I am not sure that the fine structure constant has a role either.

The criterion of the second is met by showing *through attribution* that there is a relativistic mass increase in orbit. Regarding the second, third, and fourth

---

[55] Icke, Vincent, *Force of Symmetry*, Cambridge University Press, Cambridge, © 1995, p. 219f.

[56] Icke, Vincent, op. cit. p.270.

reasons why unification is not hopeless is the use of one coupling constant with a moving set of indices, N, that can be made to denote the sum of the energy of the orbiting particles of the quark as they become relativistic. The attribute equations demonstrate a *plausible* type of symmetry applicable to the particles of both the weak and strong force. Note that a positrino has six thirds of positive charge.

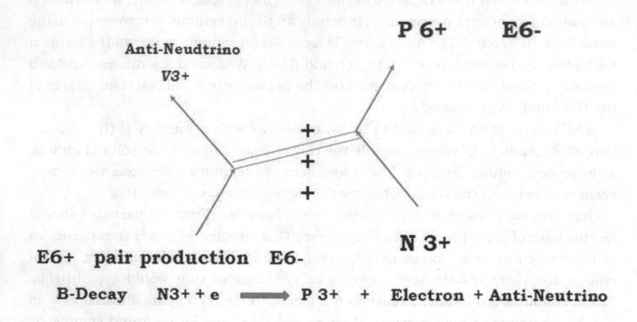

The weak force gets its name from the relatively long decay times of its particles. Various particles accompany beta emission or decay of the neutron, as is shown in Figure 6.2. Electron capture is the reverse process. This is shown in Figure 6.3. The unique difference I present in these Feynman type diagrams is the pair production and the striping of three positrinos from an electron allowing, in electron capture, the neutron for example, which gains three positrinos and become a proton while the stripped positron loses more energy and become an anti-neutrino. In B-decay the electron comes not from the neutron but again from pair production. The electron is noted in the usual in β-decay but the positron is not mentioned in electron capture.

The possibility exists that a process of pair production to form W^+ and W^- that they could momentarily orbit form W^0 and annihilate. More than one scenario delineating weak force decay exists. We could define a system in terms of the neutron and proton system of Figure 6.2 or 6-3. In defining this system of weak force decay, the particles $W^\pm$ and the W^0 and the pions $\pi^\pm$, π^0, the neutrinos v_μ, v_τ, and v_ε. It applies also to the muons $\mu^\pm$ as well; and, all their anti-particles may be included. Gravity, in the unification of the weak force, constitutes undoubtedly less than 0.00001 percent of the field because charge remains at a fractional third but the mass of the electron declines sharply. Figure 6.1 indicates the proton system of three quarks, two up each with three positrino thirds, and one down with no positrions in orbit. A neutron would have three less positrinos. The positrinos affect nuclear bonding forming single, double, and triple bonds as do chemicals. A positrino has mass of $m_e/3$ which is a third the mass of an electron. Relativistic mass *does not* gravitate. So in effect the six positrinos orbit three down quark for an overall positive charge of a proton. Shed three positrinos and you have a neutron.

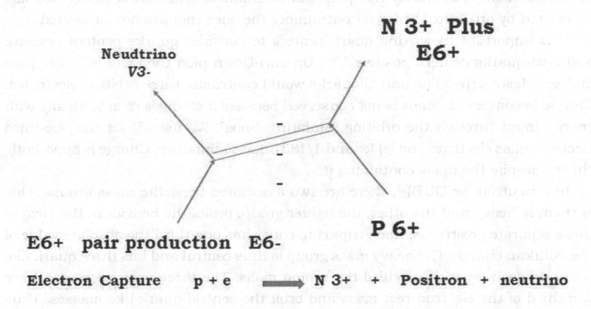

N 3+ Plus
E6+

Neudtrino
V3-

P 6+

E6+ pair production E6-

Electron Capture   p + e ➡ N 3+ + Positron + neutrino

Similarly, symmetries have internal symmetries for the orbiting system of an electron, first the charges must be reversed and the central mass is nominally one electron mass. Thus the electron is cast as a miniature anti-proton with electrinos that are 1/1868.15 the mass of a positrino. By changing an up quarkino to down we generate a miniature anti-neutrino. The mechanism that I suggest for neutron decay starts from the lone formation of a W⁻. We will use DUD to represent the central charged quarks of q/3 in the neutron, and similarly use dud to represent those quarkino centers of the neutrino. Three electrinos orbit the one-third of the central electron mass of a u-quarkino. Three electrinos orbit the one-third central electron mass of an anti-quarkino. Quark and quarkino centers are negative, while the anti-quark and anti-quarkino centers are positive. Six orbiting q/3 "inos" furnish the correct charge. The D or d centers provide no orbiting particles to UDU or dud constructs according to whether an electron (lower case) or proton or neutron (upper case) is used. One consequence is that any of the U-D, or u-d, quark centers destroy each other if one is an anti-quark. The role and nature of spin in the strong and the weak force have not been investigated.

One might question whether or not the attribute N plays the role of a hidden variable. Early classical hidden variable theories usually destroyed symmetry or implied action at a distance. By one definition, a hidden variable theory may also rationalize the behavior of a system by using the variables that are often inaccessible to current styles of experimental measurement. At a later time these variables may not seem to be as completely hidden when technology of measurement advances. The best example of a successful hidden variable theory is the chemical mole. We should note here that we calculate radii so small that they are not measurable by current techniques. If attributes are deemed to constitute a hidden variable theory, then its development is definitely not quantum mechanical. The classical mechanisms supported by attribute theory far outnumber the ones that are not supported.

It is important in writing quark centers to consider quarks centers negative and anti-quarks centers positive. The Up anti-Down pion therefore is a zero pion and self-destructive. The anti-U quarks would contribute three orbiting electrinos. Charge is conserved. Mass is not conserved because it changes relativistically with energy input through the orbiting miniature "inos". We use (E) for the one-third electron mass electrino and (e) for and 1/1837 mass miniature. Charge is apparently charge despite the mass containing it.

In a neutron, or DUDP[3], there are two associated tripartite mass groups. One of them is heavy and the other, the lighter group, orbits the heavier in the form of three separate positrinos. Each tripartite entity has one-third the absolute value of the Millikan charge. The heavy mass group is thus central and has three quark like particles each having one-third the proton mass. The three positrinos each have one third of the electron rest mass and orbit the central quark-like masses. Thus the particles orbiting the neutron are three, (six in the case of the proton) and are positrinos. Their anti-particles have orbiting electrinos. Each up quark supplies three positive positrino masses that orbit the negative center and the total charge on the sole entire up quark is positive 2q/3. Each stand-alone down quark supplies no positrinos to the orbit at all so it has an overall charge of negative q/3. Consider the neutron as three negative q/3, and a central quark mass that has three q/3 positrinos in orbit. The neutrino is a dudp3 with rest mass less than 1/1836 Mev of the electron. A flip of a down quark to an up quark requires three more positrinos that come from the positron of an electron positron pair. This generates a proton and an electron but no neutrino. The neutrino could appear through production of a neutrino anti-neutrino pair. Pair production, ostensibly, usually isn't supposed to enter into the mechanism of electron capture, as in Figure 6.3. Could this explain the negative results of the Los Alamos neutrino mass experiment?[57] The electron furnishes three electrinos that encircle the anti-neutrino to form a miniaturized anti-proton. The electron from W- decay could provide three of the six electrinos orbiting the nucleus of the anti-ν that appears.

---

[57] Glanz, James, Claim of Neutrino Mass Detection, Science, Vol. 269, Sept. 22, 1995. p. 1671.

I will describe one plausible mechanism to formulate W⁻ that preserves the two quark zero pion hypothesis. It may not fulfill all subsequent criteria. Assume that pair production of an up quark and an anti-up quark also enters into the decay. The up quark replaces one down quark of the neutron. This will result in a heavy particle, possibly the W⁻, which carries the energy of a down and an anti-up quark at its center. Being of opposite charge they could conceivably annihilate each other. However, there must be a mechanism to explain the transition to an electron without the W⁻particle.

Decay of a neutron to a proton requires only the addition of a positron of three positrinos from pair production. That would convert one down quark to an up quark, yield the required proton, and leave the electron of the pair produced. Now we need the neutrino pair production to provide an anti-neutrino for the three "inos" of the electron to orbit. This model of the electron has six relativistic electrinos in orbit about a trio of positive centers with an anti-d u d center. It has the charge of an electron and constitutes the decaying W⁻, the t-, p-, and m-, particles. They are just differing relativistic forms of the same symmetry. Such a relativistic system can decay to the mass of the electron. The electron is now a composite symmetry with central quarkino rest masses that are a factor some (1/1836.15)/3 less than the quark mass of the neutron.

We consider the anti-neutrino to be a symmetry system just as the anti-neutron is. Decay of the associated neutrino ν particle produces finally the ue particle as the orbiting positrinos lose relativistic mass. Data indicate that at the electron state the three orbiting positrinos of the electron neutrino join at a proton numerical attribute of 50.62878.

Implementing the Model

The equations of attributes of the strong and weak forces are similar. The constant in the equation that defines radius should be the same for all SF and WF particles. This is because the proton and the electron are approximately the same size even if their masses are different. One uses the definition of ? in equation 6-1 and in order to calculate the radial values. One calculates the radial function from the following:

$$r = \frac{-10^{-21}\sqrt{N+6}}{\alpha Ln(\alpha)}$$

6-1

The W symmetry system represents a 93000 mev mass particle at a relativistic orbit of only slightly less than the speed of light. I selected the radial scaling constant to reflect the known size of the electron as compared to the proton. Carbon, which has a similar system of orbitals, governed the initial selection of the numerical attributes. This selection then required a slight adjustment. Computing the requisite mev mass equivalent requires knowing the frequency in orbit. Higher attributes generate more excited systems. A distance increment of 1 x 10^{-13} meters in nuclear attribute is sufficient to separate orbitals. The entire action from W to E takes place in a span of numerical attributes, N, that is within a half integer of (spin) change in attribute. It is usual to consider one orbiting particle per orbit and prorate for occupancy. Elliptical radii insure the Pauli Exclusion Principle. Any difference in attribute is out in the fifth or sixth decimal place. The important fact is that with this arbitrary assignment of increment the equation does indeed predict that a closed form converging solution exists

The remaining equations leading to the calculation of the frequency in orbit are:

$$g_e(N) = \frac{GA(1/1836.15) + KZq\nabla\phi(N)}{r_{(N)}^2} \; ; \; V_e(N) = \sqrt{2rg_e(N)}$$

6-2 a, b

Gravity is now of a negligible amount. The frequency results from the complementary modified energy time expression. We will use the central mass of the electron in all particles decaying to the electron for the orbiting mass. We obtain the relativistic factor **g** from the escape velocity.

$$f(N) = \frac{4\pi^2 \gamma m_e c V_e(N)}{h}$$

6-3 a

The interactive formalism of the weak force often includes many path diagrams of the Feynman type. There is no implication of actual experimental observations in every one of the paths that make up the Feynman sum over histories' mechanism. The Z, or Wo, is its own anti-particle.

Zero pion, Z or W-

The three remaining electrinos of the pion jump to an anti-neutrino of a pair. There could be another kind of zero pions from the inverse process if it occurs. There is also an electron and a positron pair involved in the mechanism, whether the zero pion occurs or not.

d u d p3 = ** W-', *-, *-, **

The energy of the reaction varies from 82000 to 122.17 mev. The loss in mev for the rest mass of two quarks does not account for the energy loss. Neutrino pair production is also required to form the model. The data for the decay results from equations 6-1, 6-2, and 6-3

d u d e3 → d u d e6 + ** 6-6
105.7 Mev

d u d e6 → Electron + *e 6-7
.511 Mev

Table 6.1

Accounting for Particle Mass

Iota	N Principal	Cumulative Mev	First radius x 1024	Velocity m/sec
Z	52.23918038	93000	3.1759403632	299792457.9
W±	52.23918038	82000	3.1759403632	299792457.99987
t-	52.23918	1784.00	3.1759403654	299792457.7526
p-	52.239176815	139.60	3.1759407151	299792417
m-	52.23917	105.70	3.175940977	299792387
E	52.03827	.05100	3.202264	296830808

Table 6.2

Radial Distance between Attribute Orbits

Iota	in N	in 1S-2S	in 2S-2P
Z	1 x 10-13	1.294 x 10-37	1.29 x 10-38
W±	1 x 10-13	1.294 x 10-37	1.29 x 10-38
τ-	1 a 10-13	1.294 x 10-37	1.29 x 10-38
π-	1 x 10-13	1.294 x 10-37	1.29 x 10-38
μ-	1 x 10-13	1.230 x 10-37	1.25 x 10-38
E	1 x 10-13	1.32 x 10-37	1.21 x 10-38

The Kaons and Eta particles fall in between **t-** and **p-**. They are included in Table 6.3. It may not be possible to retain the two-quark-hypothesis for zero pions at low energies because the rest mass is too high. The mass of the system is very sensitive to the change of attribute. Selection of the increment is arbitrary in going from 1S to 2P. Clearance ranges, in order of magnitude, flatly from a high value of 10-37 to 10-38 meters in going from the muon to the **Z** particle. There is .8 % overall variation in the principal radius of the particles. Generally the mev at 1S and 2S were doubled and added to two thirds of the increment at 2P to assess the final mev. Stephen Hawking declares that no theory is complete if it contains arbitrary adjustable factors that could be given any values one liked.[58] I allowed only one scale-coupling constant that reflects the size of the electron as being equal to the proton. I selected a principal attribute number that is similar to carbon mostly because the symmetry structure of the composite masses is the same. These give a radial separation for the composites of the electron that is smaller than its radius by a factor of 10^{-14}. One must interpret this as the separation of surfaces whereupon the charge resides. The Feynman diagram at the left is intended to show the plausibility of the usual

---

[58] Hawking, Stephen, Black Holes and Baby Universes, Bantum, New York,? 1993, p.51.

definition in terms of the orbiting systems that are described. Note that in Table 6.3 that the effect needed to get the relativistic velocities requires the use of attribute N's that are closely spaced

Table 6.3

Mesons, mev	N	Orbit M/sec
Vτ 70	52.23917686	299792417.8
Vµ .25	52.032	296702384.7
Ve .00002	50.62878	276119000.8
K± 493.7	52.23917933	299792454.77
Ko 497.7	52.2391793334	299792454.82
H 548.8	52.239179371159	299792455.39

The electron neutrino behaves as if it has only one particle in orbit with an anomalous rest mass fraction of 1/1836 of the electron mass. This is sufficient to predict its mass from equation 6.4. The anomaly seems to be a logical possibility of the similar mass drop from proton to electron when the neutrino mass falls significantly below the electron rest mass. All particle masses were taken as reported by Giancolli.[59] Large numbers of all kinds of particles flood the measurement chambers of the particle accelerators. They all seem to interact.[60] The glue balls spoken of may be complexes of "molecules" of "atom like" bonded decaying particles.[61] Weak force bosons do produce electrons in radiative decay as observed by events in the W⁺ channel. Analysis of pathways of decay of particles is important; but, the resonances and their stabilities are probably a function of the contents of the beam. The fact that we can specify the mev of any reasonable resonance using attributes, because of the convergent nature of the method of calculation, should be apparent. The step of decay of one particle to another constitutes a quantum jump.

The spin rotation of the electron was once calculated by Uhlenbeck and Goudsmit to be so great as to yield a mass greater than that of the neutron. It was done in order to eliminate the non-mechanical stress hypothesis wherewith Bohr had attempted to explain doublet structure.[62] Ehrenfest had it published in Die Naturwissenschaften. When they decided it was in error he told them that they were young enough to make such an error and survive. Hopefully, we now have an atomic model with an electron spinning near the speed of sound and orbiting in sub atomic domains near

---

[59] Giancolli, Douglas, Physics for Scientists and Engineers, Prentice-Hall, Englewood Cliffs, N.J.,? 1989, p. 1028.

[60] Taubes, Gary, Mastering Nature's Strong Force, *Science*, vol. 270, Dec 15, 1995, p. 1157.

[61] Zeppenfeld, Dieter, *Nature*, Vol.378, Nov. 30, 1995, p. 445.

[62] Jammer, Max, op. cit. p. 150.

the speed of light. Possibly, we have a near zero ZPF neutrino that will vie for the missing mass of the universe.

There are methods of attacking the big ideas of humankind that may themselves be attacked. One of these concerns the concept of gravity that was finalized by Newton and revamped by Einstein. Einstein himself was preceded by Maxwell and Lorentz who laid the groundwork for general relativity which Einstein himself parlayed into the celebrated theory of general relativity. John Archibald Wheeler called general relativity a theory in which matter tells spacetime how to curve and spacetime tells matter how to move.[63] One can expect that the first mention of gravity is of central mass and thereafter one hears of a moving mass, usually a planet. More importantly it is the kilogram of mass defining that field which deserves a better emphasis. *An obvious fact is that if it were not for the first reference to matter there would be no spacetime.* Also, a person must know the location of a second matter, a test mass, in reference to the central matter in order to know the magnitude of curvature. So it is an obvious fact that if it were not for some extra knowledge of, or reference to locations of matter there could be no spacetime. When this is imbedded in that master equation containing the supposed density of the universe an estimate of the inter-point travel distances should be possible. We appreciate space; but what about time? We are thus led to conclude that the movement of some test mass is still in need of a clock. We attempt to use our clock everywhere; but, everywhere we find ourselves does not always march up with our clock. In atomic theory we seem to become exempt from such knowledge of time by resorting to phase space. Why is there not a spacetime in phase-space? Why should we need a time in phase space? This is probably not possible because seemingly gauge transformations are different for smaller entities rather than large. The answer: Accept the fact that small entities have small test mass type orbiters compared to kilogram cosmic scales. For atomic scales we employ (-10) scale factor in the equation 6-8 so that the radius is in angstroms. When used in calculation is must be converted to meters...

$$r = -10\frac{\alpha Ln(\alpha)}{\sqrt{N+6}}$$

6-8

Look for the best real quantum gravity and the nature of time in chapter seven. I had hoped to define N as a possible candidate for universal time. It does reach infinity on the future timeline.

There is not a hint of the above problems if one depicts a hypothetical space time in which all movement is relegated to a hypothetical space time conditioned on the density of the universe hypothetical if it be. Without knowing actual paths,

---

[63] Hawking, Stephen; Stubbornly Persistent Illusion; Running Press. Philadelphia, PA; p I, © 2007.

only their space times under the hypothetical density it may be possibly used to extract relevantly useful data between two points whose inter-distances are well known. Surely this must be the tactic that Einstein's general relativity relies on. It is needed when calculations must be made between entities which are in slightly different cosmic scales from another.

I earlier referred to the test mass of the atom as an electron mass some ten to the minus thirty-one times smaller than the kilogram, a fact for which I am indebted to my setting of the Kepler's constant for the atom as 1836.15 G. We know that gravity attracts; and, electric and magnetic vectors attract or repel. This dual fact of repulsion as well as attraction has caused puzzlement in light of resolving gravity as being equivalent to electromagnetic theory. Actually it is the attraction part of electromagnetism that renders gravity to be an attraction. This attraction exists when a differing balance of electrons exists in relation to the *positive electrons of normal atomic orbiting mater.* The attractive effect will not conform exactly to G when these spurious electrons are attached to orbiting mater. As we have said this results in the interpretation that some weird "dark matter" exists. Nature conspires to generally present us with large scale electromagnetic attractions. It is my conviction that the universe is not participate in an outward expansion. "Dark Energy" is a fiction occasioned by the acceptance of the fact that the velocity in Hubble's law is directed outwardly. It is not! It results from the product of the field constant and the radius, $v = r\,g$. When Warren Arp, assistant to Hubble, disagreed with his boss on this fact he was forced to find employment outside the U.S. in Germany with the Max Planck Institute. There are possibly extra electrons residing on the central mass. This is the repulsion scenario of the dark energy domain. The problem would disappear if the universe were found not to be expanding. There may be areas of the sky that do appear to be plagued with "dark energy" due to spewing of electron from supernovae onto only a central mass. The expansion presumed by the velocity term in $\frac{V^2}{D^2}$ is radial and not directed outward and should not serve as a criterion.

A moving charge requires a conductor to have a magnetic force; and, a moving magnet requires a conductor. The earth and sun provide media for both actions. Thus the electrons of the earth create their own gravitational field moving in the magnetic field of the sun. Mutual movements and their content of charge constitute fields keeping them mutually moving at somewhat of a constant distance apart. Should the earth be the recipient of a boo coup of electrons its current, and its attraction would draw it closer to the sun.

It is not my contention to contest the brilliance or usable nature of general relativity. There are thirty-five different ways in which Einstein's equation can be solved. It creates its own utility. The utility of Special Relativity is unassailable! Both the atomic bomb and the measuring devices of the global satellites are more than obvious testaments to its veracity. From the standpoint of major interest, the viability of the concept of *quantum gravity* is to be questioned because it has become the salient variable in the race to reconcile the classical theory of general relativity with the non-classical quantum theory of the atom.

A quantum as such does not seem to have an existence in solar calculations. Gravitons are what are spoken of instead. When they look for gravitons in supernovae it would seem that they are only detecting a large number of photons. Even if detected such a graviton I doubt if there would be sufficient energy to move a planet in the manner of a photon hitting the atom. If a quantum number is a real viable variable in the solar system it is likely that what we designate as a graviton will be really a collection of photons. This may possibly be a crisis that can be resolved by redesigning the meaning of the word classical.

Quantum theory considers light to be not only as an electro-magnetic wave but also as a "stream" of particles, or photons which travel with the speed of light. I think that here is no evidence, but only supposition, that these "quanta" persist as either a particle or a wave inside the atom. Their energy however does persist. What else is needed? These particles are not considered classical billiard balls, but rather as quantum mechanical particles *described by a wave function spread over a finite region in phase space*. The one usable utility is that *wave type energy* is transferred to the atom. It is not possible to know exactly which atoms have taken up what energies. *This is primarily because no one has given an acceptable classical explanation heretofore*! The trouble then would be to give a shot of energy to the earth to move it out or in. Criticism is not always constructive. Would the direction of the approaching energy decide the fate of the orbing entity?

It is the importation of definite energies to a system that has utility. The question of how this is done is moot. *Photons alone are not the only means used to deliver energy*. There is an inherent uptake of energy from a spectrum tube. My attribute equation can take input in sequences of energy from either high to low. I rather *suspect* that a given atom in a given state can accept another "quantum" of energy of any value provided it is in the range of the accepted values normally radiated by that atom and is accepted below ionization. This means that the ensemble could be highly represented *correctly* by any energy such as might be represented by any of the "spectral energy" values supplied to a hydrogen spectrum tube. The probability coded by any calculated radii is undoubtedly within the expected radii. *It would be surprising should the same probability evoke the same radial change.*

If there is anything to the graviton it is going to be an inordinately large collection of photons in terms of quantum optics. I think that those who expect to observe the collision of two gravitons will have a long wait. Virtual photons presumably have wavelength and frequency dictated at their inception. According these same conditions to gravitons probably will result in frequencies values that would seem to exceed those implied by the range of frequencies found in light. Assume the graviton, like the photon, is electromagnetic. Then the burst of light from a supernova is a collection of photons and possibly as close to a graviton as will ever be found.

Therefore if such great energy is detected it is doubtful if will be interpretable in such a way as to verify the hypothesis of the graviton.

Truly, any concept in use for some long but unspecified time may be said to be classical. So far the physics community *has not considered the quantum number to be classical* when applied to the atom; nor, has the quantum number been called classical whenever, or if, it has ever been successfully or otherwise applied to general relativity. I clearly have demonstrated in chapter two that the quanta, namely the count (n) of the incoming photons, have produced a classical effect in hydrogen and helium. That unit value of an always unit photon does not include the same set of connotations of the idea of a quantum wherein momentum, $n\,h$, n = 1, 2, 3… is being conserved.

The idea of conservation of angular momentum, I believe from the de Broglie concept expressed in effect, a non-existent symmetry between the photon and the atom. The implication of this is that angular momentum is conserved between disparate geometric entities. The early success of this is that it supported Bohr's idea of stable orbits. It also inspired Schrödinger to write the wave equation. It was later realized, as I have shown, that the equation $n\,h$ = mvr results in hundreds of orbits of the electron before the matching of wavelengths could occur. The radii that were involved were undoubtedly not acceptable.

Edward MacKnnnon has a very concise article in *The American Journal of Physics in* which he goes into great detail regarding the de Broglie derivation and why it does not work.[64] The fact that Einstein came to the rescue of the de Broglie thesis is explained by saying that he, Einstein, had approached the same topic from another standpoint.

Certainly the integers were known before the number line. That we now in *quantum attributes* employ a number line and use real number integral differences to correlate radii and order integral quanta is not such a far out concept. The fact that we can produce a periodicity of principal attributes that is the reflection of the periodicity of atomic radii with atomic number is just one more convention that tends to support my theory. Indeed, we now have a classical theory of gravity and electromagnetism inside the atom as presented in chapters one, two, and three. When energy is imparted to hydrogen or helium there is no role for the quantum number n in any equation regarding the atom regardless of whatever manner.

The next most prominent facet employed by quantum theorists is the so called hidden variable approach. These were variables hidden to quantum mechanics. Quantum mechanics considers energy that is deployed in bundles. Ancient Greek peoples (Democritus in particular) held to atomistic theories of matter which rendered a small piece of matter uncutable while (Aristotle) believed matter was

---

[64] MacKinnon, Edward; de Broglie's thesis: A critical retrospective, *Am J Phys,* Vol 44/11, Nov., 1976

just highly cutable stuff. Energy is easy to follow; although, it is not certain that the energies enter the atom starting with the least and building to the greatest. For one thing in a spectrum tube, where energy is supplied by an electric field, we deal with an ensemble which often may indicate a preference for one frequency over the others depending on the voltage. In this case one may question why angular momentum should also supplied as if the excitation were by entering photons. I would like to see a breakdown of the spectrum as various voltages were to be imposed on the spectrum tube.

Consider a photon whose energy $hf = M_0 c^2$. Despite Planck's constant with its implied angular momentum, how can the photon move in essentially a straight line with virtually no angular momentum? What happens to the M_0 when the photon enters the atom? The M_0 is an illusion of the equivalence of energy and mass. If we were to recover any angular momentum of the photons that enter an atom we must conclude that an angular momentum preserving M_0 is required to be found in the atom else symmetry of angular momentum is broken. This surely is not so because if the energy came from a spectrum tube would have to provide M_0's for exiting photons which is highly unlikely. If the M_0 and the hf both enter the atom the additional equivalent of energy of its mass would double the energy that the atom would receive

We should just state that angular momentum is not conserved, a fact that quantum mechanics does not always ascribe to. It would seem that they might want to retain Planck's constant as both angular momentum and frequency that is imparted to the electron. We can debunk that because I think that all they do is impart an energy and ascribe no physical properties of the atom therefrom. If they achieve macroscopic ends that are useful then their procedures are going to be superior because I must now deal with an ensemble. For want of a reason to replace their notion of conservation of angular momentum, it may peacefully exist without any benefit or detriment. This conservation would require a mechanism to repatriate M_0's with photons when the photons are emitted.

Consider a photon leaving a star. As it travels toward earth it passes matter which becomes an increasingly greater concentration at the point of its departure by Newton's law. The photon works against this gravitational source and loses energy and thus frequency while increasing wavelength and increasing redshift. The perceived angular momentum is zero; and its energy becomes less. This happens! When the photon gives up its energy to an orbiting electron we conclude that any M_0 goes to zero; and, there is only energy left to transfer. If we were to conclude that when the energy described by Planck's constant times frequency transfers that the residual M_0 were to remain, then; the ocean is full of these M_0s. The only foggy part of the scenario without residual M_0 transfer is how the exiting photon obtains a central M_0? It does so because of the mass energy equivalence relation given by Einstein. Undoubtedly it does do by the same way when as a photon it leaves a star.

I attended a meeting of chemistry teachers in Canada where a protégé of the de Broglie Institute gave a paper in which de Broglie claimed that there were two clocks in the atom. This meant to me that there are two orbiting entities in the atom. I tried to follow this lead for several years. It always disturbed the energy balance which is more important. The conservation of angular momentum is one of the popular views of quantum mechanics; yet, when you consider that imposing some twenty frequencies on the electron as it orbits you have a real donnybrook of squiggling frequencies.

I wanted the de Broglie hypothesis to succeed in the atom because it came out of the Rydberg rearrangement. Since it cannot be associated with the expression of the unified field; but, it is connected to the solar case and the impact of it is to predict solar quantum numbers. Planck's constant must be modified for this to occur. Some combination of Avogadro's number, the speed of light, and the charge on the electron can be used to modify it. As it first occurred it was from the Rydberg rearrangement as $N\,h\,V = G\,q$ which is an equivalency that I firs used and gave real quantum numbers between 2 and 20 for the planets. There is considerable effort needed to relate these numbers to precession because of the interest in quantum gravity which I have yet to do.

Let me again describe a quark system. I can demonstrate an orbiting system for the electron that will show it to be about .52 mev as a relativistic entity. We will define the down quark as an entity having -q/3 charge. The up quark is a down quark orbited by +3 q/3 units of charge. One may now get the proton as UDU and the neutron as DUD. The proton is three down quarks orbited by +6 q/3 units of charge. One may now think of an electron as a small "anti-proton" or **u d u.** Its orbiting charge is negative. A UDUP6 represents a normal proton. Change that to UUDE6 and I mean the anti-proton. Since the central particle is always three or less down quarks, the E or P tells us the central charge be it negative or positive. For example P6 implies the central mass is negative three. If it were a P3 we would mean a neutron. Given dudP3 it would imply an anti-neutrino. The ratio of mass between a UDU and udu or a dud is 1815.6 to one. The electron becomes uduE6 and the electron neutrino is dudE3

I have questioned the use of the Lagrangian in formulating wave equations involving orbiting electrons because in orbit the potential $m\,g\,h$ and the mv^2 are one and the same. It assumes conservation of energy in free fall. I have employed a Newtonian system for which the Lagrangian does not apply. Nevertheless, it was the Lagrangian that Euler and d'Alembert used to formulate their wave equation in the 1750's. It is said that Einstein employed a Lagrangian in developing general relativity. If so he surely must have a variable attributable to the Poisson charge density in his equation. My Poisson stems from the vector potential. I showed that the Poisson portion of my Bessel equation could be derived from the radial separation of the Schrödinger equation which was formulated with regard to a

Lagrangian. This could be expressed by a quadratic. My quadratic resulted from the relation between the vector and the time average portion of the Bessel equation when it was set equal to zero. The Schrödinger radial equations contained a second derivative which I could set equal to zero because I assumed that it had to be zero vector in the time average direction. Only the Taylor series and its first derivative with respect to ? plus the time average portion occurred in my derivation of the Bessel equation. This allowed me to solve the parametric equations for the Bohr radius as I indicated in *chapter two which was a first grade serendipity.*

If from the above we postulate that every position has an associated electron density then there is no bar to asserting that the use of the Poisson. In attribute mechanics is no bar to their use.

Chapter Seven

What Time is it?

One of my father's favorite questions was "What time tis it?" He introduced me to magnetism of iron fillings in his shop long before I became of school age. He explained to me that only four persons understood Einstein's general relativity in those days. The four men, I later found, were Einstein, Eddington and two other fellows, names unknown. My father taught the men's Bible class on Sundays for fifteen years. In his younger years he was a Fire Boss in the coal mines. During WWII he worked in a 3-foot seam of coal. I wonder what his reaction today would be if my father and I were to discuss reality questions. The problem with reality is that everyone is presented a *set of values* obtained from their view of the world about them. My missionary mother taught elementary school before I was born. She also taught the women's Sunday school class. It stretches ones sensibilities when something one has learned early in their life comes into question in their later life by someone with a different value system.

Early time and future time were two times feared by earliest of people. Both past and future times embody infinity. Infinity was the concept that was really feared. Fear happens because both end of the infinity of time are implied when questions are asked like "where did we come from", and "where are we going"? In addition there is the fleeting "now", a type of zero always with a place on the totality of the number line of time. Strangely enough zero was shunned in the west until its utility in mathematics was demonstrated by mathematicians of India. The Romans had no zero. We mostly deal in intervals of time for there is little time for any other kind.

Time was said by Sir Isaac Newton to flow eternally, unchanged forever and supposedly so disproved by Einstein's general relativity. Can we now resurrect Newton's conjecture? In reality each orbital system is its own clock and, barring a novae of the central mass of a Sun, *does flow* eternally. That eternal time may just be a number of repeating intervals until the central mass goes is slight comfort.

So in actuality all our historical times are really a difference in time. One proof of the big bang is that if it had not happened we could not have reached the time where we are now. We define our second as an interval. One whose second depended on Jupiter's travel about the sun would have a different count of the number back to the infinity of time; but the big bang's time of occurrence would not altered. Our interval, the second of earth, can be transformed to any planet. Is there any place where we are not able to be in orbit?

The trouble is that we don't have two clocks that agree on their "absolute time interval" when an event is viewed from their twin perspectives. We can assert a certain differential character of Newton's idea of the eternal flow of time when we find a number for the rotation of another planet and relate it to our own. We could assume 100 meters were available for the dash on this other planet and calculate a meaningless velocity that would occur if our interval of time were converted to theirs. Time is not a number on God's clock unless he has selected an interval for the second.

Any time interval in the universe bears a ratio to our selected interval of time. Every planet rotating about any Sun has an N like quantum gravity. Quantum gravity N will not be an attribute which we earlier defined as a number between 36 and infinity; because, it undergoes a change in character when applied to the solar system. Our year being 365.26 days we need to find an N for a planet that takes 400 days to complete an orbit. We can look for such a planet or just assume that we have timed a planet with such a period. Since N in the solar system cannot be considered an attribute any longer, I shall rename this N to be a de Broglie solar quantum number.

$$NhV = 2\pi qG \qquad \text{7-1:}$$

$$V^2 = \frac{GM}{r} \qquad \text{7-2}$$

$$\frac{N_{Earth}}{N_{Exo\,P}} = \frac{365.26}{T_{orbit}} da \qquad \text{7-3}$$

Assume that an earth person observes or postulates a planet which takes T=400 earth days to complete an orbit. Can one find the central mass of that planets orbit? By equation 7-3 we can calculate the quantum number of this planet outside of our solar system. . Using this new quantum number we can calculate a velocity for the planet by equation 7-1. The radius is then the product the now known velocity and time in days converted to seconds. Armed with the radius and the velocity, the central mass of the planet can be found by equation 7-2. If the days it took for the planet to orbit were 400 then the New N is 1.86 assuming earth is 1.7 by equation 7-1. Putting 400 days into seconds we find that the velocity is 5.44×10^4 m/ sec; and, the radius is 1.88×10^{12} meters. Finally by equation 7.2 the central mass of our new planet is 8.36×10^{31} kg. This works because we employed earth units. If we used their second the distance unit would not be meaningful.

Using equation 7-1, this time with *mean orbital velocities*, I find the series of quantum numbers for our nine planets to be (1.08, 1.45, 1.70, 2.08, 3.48, 5.26, 7.44, 9.35, and 10.69). N is probably not the candidate for Newtonian time that one would expect; but, every rotation time relative to earth has a different but related real quantum number with a unique central mass. We must select a unit if time is going to be numbered. At least we have the relation now between our second and any second that will ever be or has been. Our clock was originally keyed to the circumnavigation of the earth about the sun. Reflecting on equations 7-1 to 7-3 we find that the quantum numbers of all planets of our solar system will all give the central mass of our Sun. Currently we define a second as some nine billion vibrations of a cesium 133 atom. On any other planet the time of its passage around its sun would require revision of the factor multiplying the vibrations of cesium in order to define the second on that planet if we insist on our definition of time. Perhaps there is another dimension to time besides Einstein's fourth dimension of time.

I am amused at the creationists who insist that they can pin creation of the earth to between say six to eight thousand years ago. They roundly condemn anyone who counters their claim using geology or the formation and life of stars. This is worthy of note because they are trying to influence the educational curriculum. They also wish still years after the Scopes trial to bash evolution because they do not understand it. There is nothing in the good book to mention evolution. But, if it is the book that describes our progress towards knowing God, then it evolved and translations still evolve. There nothing in the good book to contradict evolution. Creationists also claim that the stars were created along with the sun several days after light was created. They waste money trying to reach young minds because they do not understand that science is just knowledge and they do not accept scientific proof. Science has always respected the early accounts as truth based on the knowledge known at the time. Scientists such as Kepler and Galileo, and Newton were devout men. Darwin was the son of a cleric. His discovery of evolution is a word that few people realize is a change that occurs in germs and the viruses for their self-preservation. Many ignore the changes in plants for the better and think only that someone is telling them that mankind descended from apes, god forbid. The knowledge that environment brings about change is the knowledge of which they for the most part are very willing to accept. No world order will change environment so as to be the same for all. There is always the eternal hope that individuals can change their environment to their own betterment in the eternity of time that we have left.

Pick a length of time to call a second then there is the *concept of a clock that is omnipresent* and has recorded time from day one. None of the niceties of the above observations regarding time really speak to our current difficulties with time; but, thanks to GPS our system is in excellent shape. So, anyone can pick an observer and arm him with a clock. Thought experiments are low budget and even with that most people will say that they just don't understand science. There will be a reluctance for many who read this book to accept it. Why should the number in one of the equations be an eight instead of a seven? Because the rock bottom start is not treated from Maxwell's equations or from d' Alembert's equation why should electron density be what it is? Also, why must electron density change when the attribute changes? If it appears to work would it be more accurate if the parametric equations were changed slightly?

Quantization came to imply the entrance of angular momentum but there is no angular momentum to enter since there is no mass for the photon. Quantum mechanics ostensibly still advances the theory that angular momentum is conserved which is fallacious. Had atomic radii been calculable there might not have been need to resort to conservation of momentum or uncertainty. Nevertheless, the lack of a viable equation for radius in the atom was also the factor that would been have *least likely* to have been found. I know because of how unlikely it was that I found it. How anyone could possibly have hoped to *measure* the location of an electron from its central mass I do not know, let alone the position of an electron just hit by an electron in an ensemble. The remoteness of this possibility is partly responsible for uncertainty, I'm sure. The position that they really were measuring was an electron external to an atom in a cloud chamber which resisted the uncertainty of its velocity and vice versa. Again we find this is the specter of wave-particle duality transferred to the atom. Much has been made of the Uncertainty Principle perhaps much more for it than in opposition to it. Consider again de Broglie's hypothesis. Could he have found it as I did by rearranging the Rydberg equation? In a final hope for the spirit of the de Broglie hypothesis, I can say that there is only one possibility left for it. First it cannot apply to the atom as it is inconsistent with chapter two; although it does support my rearrangement of the Rydberg equation *if n is restrained to be one.* Left is the possibility that there it characterizes a non-integral quantum gravity. This chapter has confirmed this notion.

A closer examination of one portion of my derivation of the Rydberg arrangement shown in equation 7-4 shows a radical containing two velocities one of which is the speed of light. I reasoned that this was a velocity less than the speed of light. It was obviously not of the same magnitude of the velocity on the other side. There is an added factor of 2π which I completed so that Planck's constant could be divided by 2π. I also dropped the q in equation 7-5. This LHS of the equations of chapter one that was equated to G is:

$$G = \frac{nh\sqrt{cv'}}{2\pi qMm} = \frac{nhv}{2\pi qMm} \qquad \text{7-4}$$

117

From which: $$\frac{GM}{r} = \frac{n\hbar v}{mr} = v^2$$ resulting in 7-5

$$n\hbar = mvr \qquad\qquad 7\text{-}6$$

$$\sum_1^n nhf = m_e(\Delta v)^2 \qquad\qquad 7\text{-}7$$

This was a neat way to derive de Broglie's hypothesis. Equation 7-7 refers to the change in velocity.

However each photon entering is entering as a unit of one photon instead of a counting number n that is building the angular momentum inherent in the units of Planck's constant so that n appears as a variable in the treatment inside the atom. So angular momentum never builds up within the atom; but, the energy of successive photons does build up. No counting number, n, accumulates within any mathematical construct of the atom. So any reference to a quantum theory with n= 1, 2, 3, ... wherein n refers to anything but the outer amounts of successive unit quanta in energy, not angular momenta, entering the atom. This goes against de Broglie's hypothesis in the atom as the value of n nowhere is encountered in any calculation occurring within the atom. Equation 7-4 is the correct interpretation of de Broglie's hypothesis although this energy results in a radius that becomes smaller with entering photons. Actually the de Broglie hypothesis as in equation 7-3 is now currently disavowed for other reasons. Where ever Planck's constant appears in the Schrödinger equation it has no basis in fact as a number within the atom because no multiple of n, or even Planck's constant itself, cannot enter into the internally described energetics of the atom. The name quantum thus refers solely to photon so that half of the intent of the name quantum theory is a misnomer.

We now have the verb to quantize which is applied indiscriminately to any misunderstood process. Nevertheless, any *useful information extracted* by means of Schrödinger's equation is apparently exactly that, namely useful. Strangely, since there is no quantum number n in the equations of the atom there is no mathematical need in the atom for the *nth plus one photon in the list to follow or precede the nth photon*. This is not to say that it can't

I have thus puzzled over the meaning of de Broglie's hypothesis as pertains to the atom. I must conclude that it expresses something *inherent in the Rydberg equation* namely *that a single photon at a time* is imparting energy to an electron. While I used it to find a real (non-integral) quantum number for the planets, I at one time had to construct a macroscopic version of Planck's constant in order to do so. Consider now that it is attached to only the solar case and not the atomic case. It will survive as the de Broglie real numbered quantum number.

Should the angle which the electron mass is contacted by the photon be such that only a portion of the energy is imparted to the electron then not all its energy will go to changing the radius of the electron. This is antithetical to there being only certain acceptable jumps. Ostensibly the entire atom has to receive some kinetic energy as a whole. Actually the temperature of helium when it is fully ionized is elevated to some 5000 °F which means that excitation of the electron in the atom is attendant with temperature rise and thus some extra mechanism promotes elevated temperature. When energy is supplied via a voltage one considers only energy radiated as energy taken up; although, the gas is carrying sufficient current to no doubt elevate the temperature as well. I doubt if anyone has tried to excite helium with a cocktail of its spectra to see how it would reradiate. Originally, Niels Bohr depicted hydrogen with negative energies which was difficult to comprehend.

The use of the Lagrangian is antithetical to Newtonian theory. Whether Einstein in construction a correction to Newton's energy found justification to include a Lagrangian is open to question. Friedman did, I think, employ a Lagrangian in his three dimensional characterization of general relativity. My version of a unified field challenges this quantum electrodynamic development and questions the nature of its usage in the Standard Model as well. Currently, since both employ the Lagrangian the reason that they do not reconcile is largely due to gauge theory. If gauge invariance cannot be sustained because of size it is because of size it is because phase space is devoid of a length dimension. This means that there is no bar to the compatibility of general relativity and attribute mechanics. What matters is that now under attribute mechanics the gauge problem is purely a matter of scale. That is the change of test particle is the only alteration but otherwise both systems employ the Newtonian system. This should guarantee that attribute mechanics is compatible with general relativity even if the latter uses the Lagrangian. I have speculated that the Lagrangian as used by Einstein went to inclusion of an electron density that has become known by another name so as to lose significance. The invariances that caused the failure of the compatibility test with general relativity are no longer present because use of the Schrödinger equation as such is no longer indicated.

Heretofore both GR and QM, presumably, used the Lagrangian. Bothe really applied to universes. It was a gauge equivalency that caused the inconsistency of the two. QM related to universes of atoms. GR related to universes of solar systems. The QM or atomic case was usually for like atomic universes. GR was intended for unlike multiple solar universes. One notes that the Lagrangian was fashioned outside phase space. Its use in phase space can either be said to not apply because there is no linear dimension there with only additive energies. The use of the Lagrangian was obvious in the QM atomic case while it was largely hidden in the GM solar case. A modest case where the distance was between adjacent solar systems the imbedding of the Lagrangian into the expression involving the dual centers of mass would satisfy the "dropping in" requirement

While I do not pretend to be a master or even a serious student of General Relativity I can speculate about it. I am unable to follow theory though the tensor medium. Take a half dozen solar systems, a galaxy or two galaxies. Assume the average density of the system. The problem is to navigate between two places that are more or less in each other's line of sight. It is said that you must know where the masses are. The geodesic route might depend on traversing a route best monitored by the center of masses of the complex but perturbed by local masses. I would orbit the complex to a point where I could "drop in" with the aid of the gravity of the target.

Any graviton is probably a collection of perhaps photons or energetic electrons ejected by a supernovae. If gravitation is also a magnetic force, and that was spoken to in chapter one, there is no need of any other agency, such as the graviton, which were suggested as being comparable to the virtual photons postulated in atoms that are also part of the Standard Model hypothesis. In short, if energy is supplied by a photon to an atom which causes it to increase in velocity and orbit at a lower radius then of what service would a virtual photon be rendering? Would it push the electron out? No, it is said to serve to keep the electron in orbit; but, I presume only if the velocity of escape is not exceeded. Consider now that old usages of electric fields had a persistent Newtonian flavor. Would it force it out? Something would, when the velocity of escape is exceeded.

Consider a ball spun and restrained by a weight on a string through a slotted tube when twirled faster. The ball goes into greater orbit unless further restrained. Conversely if the restraint were decreased the velocity in orbit would have to be increased to keep it in orbit. The analogy of the ball on a string is not entirely exact. The analogy of the spacecraft in orbit is exact. Why do we not hear of virtual photons keeping our spacecraft in orbit?

Time as we have always known of it depends numerically for its unit value of one year taken to complete one revolution of the earth about the sun. We subdivide the year into seconds which tells us the distance a point on the trajectory is traversed in that unit of time. Since ancient time scientists have found devices of wheels, cogs, falling water, and atomic devices to mimic the second of time involved in the daily duty of the earth about the sun. Yet the beginning and end of the time number line are inferred but not exactly known.

I repeat myself often. As my Brooklyn Landlady used to always repeat when speaking of a parking violation to the judge; "Guilty your honor, but with an excuse!"! It is said that if previous time were definitely infinite we would not yet be where we are today.[65] I earlier noted that this one argument for so called beginning of the universe called the big bang. The idea that someone could have initiated a count of years since any bang, is not possible. If the year depended on the circuit

---

[65] E. Williams, J. Fuller, and H. Hill, "New experimental test of Coulomb's law" Phys.Rev. Lett 26, 721-724 (1971).

of an electron in hydrogen we might be older in number but no wiser. If we knew the year in which the universe had been created that knowledge would not improve our current state of knowledge one whit as to when the earth was created. And if we knew in the years ahead the time after which creation might no longer be sustained, we could do little about it. Galileo, Kepler, and Newton were all deeply religious; yet, they lived during the era in which science was undertaking to extract the philosophical nature of where we came from and where we are going. This was an occupation which had previously always been a function enjoyed by the clergy. Some say science today is to mankind what theology was prior to the enlightenment. Whether we like it, disagree mildly, or fight it, we are on a road toward a resolution of the enmity between theology and science. Theology is very adamant, particularly with regard to creation. There is there an element which will resist any science but that which will cure them, even if it is of something they do not know they have. Many would stand on street-corners saying, if asked, that they never understood algebra.

Without discussing the evolution of the philosophy of reality from ancient times, which I cannot. I will just assert that there are two classes of reality. There are those visible realities that all can see. There are also the multitudes of differing realities that exist in all the multiplicities of minds of mankind. The former kind are geographical. They change both with population shifts and technology. The latter kind are conceptual which are more difficult to ameliorate. The very first class that I attended in college was senior level course in philosophy. One of the later and most recent courses was a course in philosophy by the same Professor. The difference was extraordinary.

No one is privy to the scope of realities existing in the minds of mankind. Some of the best aspects of it are found in places where little education is available; and, there are other places where there is a mindset that no amount of education will seemingly cure. Examples of the both case abound. Education is painful. Letting go of a belief or procedure can also be financially costly. There is also indifference. An example that suggests itself is the case of the person admonished for disturbing the sermon of the Parson. When asked if her were going to that upper or to that lower region upon his demise, he rejoined that "it didn't matter because I have friends in both places".

Religion supplanted the Roman Empire because the Empire relied on religion to keep it intact. Religion is now relying *selectively* on science to keep intact its power base which is doing just what the Roman Empire did. That is religion brands certain areas of science as being heretical. Only if, or when, the curriculum is 100 % correct and available for all throughout their lifetimes will education be indeed a universally sought for offering. Only then will man ever subscribe to a universal creed. Such a world, unlikely as it may be, is probably not as desirable as everyone might wish for. For those who seek to enhance their knowledge there is no certainty that education is now or will in the future be available for either the things we might wish to know or to search for things which we know not of but want. Mark Twain has said "that we are all ignorant; but, on different subjects". According to Confucius, "When you know what you know, and know what you don't know, then you are wise indeed!" All are willing to take the blessings of science while at the same time ignoring its lessons.

Uncertainty came about because an enigma regarding an elliptical fit to a linear oscillator in phase space, namely the Schrödinger equation, which allowed only certain ellipses.[66] The Poisson portion of my equation is related to the Schrödinger equation obviates uncertainty serendipitously. Einstein did not believe in uncertainty. Contrary to common belief that he didn't prove that gravity and electromagnetic phenomena were the same, he did so indicate that they would be if in a Poincare universe with a constant that could have been the Poisson. If the Lagrangian did not result in a Poisson correction to general relativity then I must admit that I am skeptical regarding the utility of the Lagrangian in general relativity because of three reasons. First the units of each term of that equation are those of the Hubble constant which requires only the mass density term to be valid in any solar system. The remaining terms have to be due to the use of the Lagrangian. Secondly I think that the applicability of the equation to *multiple* galaxies is not aided by resorting to the Lagrangian even though it might be useful in certain regions between two galaxies. I cannot imagine any tensor that could tackle such ingenuous placement for more than two galaxies. Thirdly, the ad hoc cosmic constant, unless Poisson dependent, and Hubble's conjecture of the outward velocities, are *probably* found only in certain areas of the sky. This is not a fix for both dark matter and dark energy.

The development of the concepts for calculation expressed in this book has happened over fifty years. You would not recognize the original derivation of the Bessel function in cylindrical coordinates in chapter one because it was done for a different reason and with different variables yet evidently embracing a similar process. In that derivation there were not one but two alpha variables representing two closely similar distribution coefficients that stayed the same while N varied with the number of stages of separation required. As later in finding the algorithm for the radius of hydrogen I wrote the solution of alpha as an exponential because I found two Taylor series in the original derivation that could be so expressed and utilized a variable rho. I found the radial formula which is a parametric equations by inverting rho in alpha. Rho does not enter into the usable equations. There is no concept of the radial or Poisson variable in solvent extraction. Rho is the bridge between the Taylor series, the parametric equations, and the associated Legendre equation.

---

[66] Müller-Kirsten, Harold J. W.., Introduction to Quantum Mechanics, World Scientific, ©2006

Only after years of use did I realize that I could define a unified field. The unified field was viable only if the electron mass was used as a test particle. Later still I found that the unified field could be derived from the Rydberg equation. After that I was able to replicate the first two terms of the Bessel function from the associated Legendre equation. The first two terms of the Bessel equation constitute the vector potential which can be converted to the scalar potential. The scalar potential is the Poisson. The last two terms of the Bessel equation are equal to zero for all N but define alpha in terms of N which is the key to finding the scalar potential from the vector potential. It was from the area of the last two terms of the Poisson that Schrödinger did the time dependent development. By these terns being zero in the Bessel equation we have the time independent case from which a jump in energy seems to smoothly change radius. Al this has been presented earlier; but, repetition and presentation in a more compact form may aid the reader. Refer to Figure 3-1 which is a description of the d' Alembertian or the Bessel equation or the Poisson, depending on which part you perceive.

I believed that spectra required additive radii; and, for some time I did so because when the electron ionized and escaped the atom it was already out and away. Also, the algorithm for the radii of the bulk of the atoms required radii that added to the radii of lower orbitals. Instead later that I found that one must allow an *attribute to change* in order to find a change in radius occasioned by the energy of a spectrum; and, from this find the absolute lower radius brought about by the change. I must iterate an attribute for the second radius to find the Poisson limits of the second radius in calculating eccentricities. In contrast to obtain greater and greater orbital radii of atoms beyond helium I added orbital radii by compiling matrices of orbitals and adding their radii instead of subtracting.

The splitting of the spectra due to electric and magnetic fields has caused no problem. It is time to address the particle zoo as it is called. Those particles that appeared as "vees" in a bubble chamber constitute the major portion of that variety that go by the name of mesons. They are composed of two quarks. The three quark particles are the proton and antiproton. Usually pions are said to contain an up, down or anti-up, and anti-down particles with occasional strange and top and their antiquarks included. I will call upon Occam's razor to ignore the strange and top quarks of the strong force for now because I believe that they are just orbital systems having a greater energy simply represented by orbiting q/3 charges traveling at a higher velocity representing a velocity indicating a greater relativistic state of mass.

In order to initiate additional study of the quarks of protons and electrons let us consider the weak force example wherein the beta-decay of a neutron takes place resulting in a proton and an antineutrino. The arrows are not exactly easy to explain but the salient feature is that a downquark goes into and upquark and a negative W boson decays into and electron and an anti-neutrino. The products are not in question; but, this mechanism is one that can now be improved on without damage to the concept of the gauge boson. If one assumes that pair production occurs and an electron and an anti-electron are produced we can describe the loss of three small +q/3's from the anti- electron that transfer as a W+ (arrow reversed on the W) so as to orbit the UDD neutron thereby changing it from a neutron (three negative's orbited by three positives) into a proton (three negative's orbited by six positives). One should complete this by reversing the direction of the arrow on the anti-neutrino as we did on the arrow of the W- (changing it to a W+ boson); and, keep the electron coming out as it is. The essence of this procedure is to add structure to the quarks and to the "newly designated" particles. There are three kinds of gauge bosons. Photons carry the electromagnetic reaction, the W+, W- and Z bosons carry the weak force interactions. And the gluons, the carrier of the strong interactions. Some of the equations in the latter portion of chapter six were taken from my first book. I hope that this leaves room for the Higgs particle.

I also have a new and novel explanation of the decay of the mesons by the weak force that conserves charge. Muons are simply excited electrons. An electron traveling a 299788951 m/sec will have 206.7 times the mass of an electron as shown in the table below. We can configure the π+ meson as a negative q/3 orbited by four positive q/3 charges. Consider pair production where an electron muon loses three electrons zapping the pion with the negative and now lone pos0itive orbiter zap each other. Now we can again have a weak reaction whereby a positive muon and anti-muon the anti-muon result in the positron and the electron neutrino with only the μ+ and the v_μ left.

To achieve a W+ three positive q/3 charges destroy three negatives of the anti-muon leaving it the positive muon antineutrino with the positive muon intact.

	Kg	Mev	Factor	Velocity
Electron	9.10938E-31	0.51099891	1	1
Muon	1.88353E-28	105.658486	206.768516	299788951.9
Tau	3.16747E-27	1776.82	3477.15027	299792445.6

I am assuming that the positive muon is really an anti-electron that is at a higher energy than a positron. Those six positive orbiters remain intact; but, the six negative orbiters are in the other half of a pair production of muons similar to electron positron pair production and are zapped by the three "positrinos" of the W+ emanating from the pion+. Since the mesons are all similar having either one positive center orbited by four negatives or one negative center orbited by four positives we can have either a W+ or W- boson emanating from the pair production center to the meson with the creation of either an anti-muon or muon neutrino. Refer to the energy diagram above and note that the energy of the muon is about 75% of the energy of the pion which was zapped apparently at the expense of an electron while the neutrino does little or nothing toward making up the difference.

The W- bosons must account for loss of energy of the reactants. W and Z bosons correspond (roughly) to the three generators of SU (3) in weak gage theory.

The discovery of the W and Z bosons was one of major success stories at CERN. The Z boson is a neutral current. In a huge bubble chamber the tracks of a few electrons suddenly started to move, ostensibly of their own accord, and were photographed. This was interpreted as a neutrino interacting with the electron by the exchange of momentum imparted to the electron through the interaction of what was called the Z boson. The neutrino is otherwise undetectable. Thus the only observable effect of a neutrino's presence is the momentum it imparts to the electron through the exchange current, the Z boson.

What is left to show is the calculation of the orbital systems required to produce the more energetic versions of the quarks, mesons and electrons going to muons and tau particles. Besides the original *Up, Down* quarks we have the most probably relativistic *Bottom, Top* pairs as well as the *Charmed,* and *Top* quarks to consider as we check out some of the remaining particles of the zoo. Because the closer the orbiting particle to the speed of light the greater the mass, we can always find a velocity of orbit that causes the mev of the particle to agree with the mev of a particular inhabitant of the "zoo". In this manner the *time* of orbit (i.e. its velocity) fixes the mass of the particle. It is thus seen that *whatever fixes our time* that we call the "year" of orbit, also fixes the mass of the particle. I will provide examples of this effect shortly as I describe why the neutron and proton have the masses that they do.

For technical reasons involving gauge invariance the gauge bosons are described mathematically by field equations for massless particles. Therefore, at a naïve theoretical level all gauge bosons are required to be massless, and the forces that they describe are required to be long-ranged. The conflict between this idea and experimental evidence is that the weak interaction has a very short range and requires further theoretical insight provided by the Higgs mechanism. The Higgs particle has recently been observed and the mechanism is considered to be complete. Thereby the W and Z bosons of the electroweak interaction gain mass, coupling by providing ripple to the Higgs field and undergoing spontaneous symmetry breaking. The remaining gauge boson, the photon, remains massless. Assuming the graviton is massless it is futile to search for it as it is very probably a collection of photons. This is enigmatic because it implies the supposed virtual photons of the atoms are not virtual. You paid your money so you can take your choice

I now can amplify the role of the gluons when they mediate the strong force. Usually it is generally conceded that no one has observed a lone quark. Therefore what is designated as being properly expressed as an *Up* quark going to a *Down* quark or better still a proton going to a neutron, is a strong force reversal of a weak force. Gauge bosons are the quanta of the gauge fields. There are as many gauge bosons as there are generators of a gauge field. In quantum electrodynamics the gauge group is U (1). In this case, there is only one gauge boson. In quantum chromodynamics the group SU (3) has eight generators requiring eight gluons. When we have pair production and for simplicity let us assume that the orbiters are each positive q/3, the down quark can go to an upquark by acquiring three positive q/3. So given a proton with two upquarks going to a neutron, can one up quark go to a down quark creating a neutron from the proton by losing three of its six +1/3 orbiting charges? Yes. Pair production of an electron positron pair can provide either three negative q's with emanation of an antineutrino or three positive q's with the emanation of a neutrino. The production of the former is the beta decay and the latter would be the reverse production of a proton from a neutron. I don't think that the latter reaction to produce a proton has been observed. So we can have one up quark of a proton going to a down quark which is what happens in weak force beta decay. Below in Table 7-4 are the details of the masses of protons and neutrons.

The rest mass of an orbiter was calculated from the formula:

$$6(1.037822946772)R + 3(1.040268772)R = 2.3055 x 10^{-31} \qquad 7\text{-}8$$

Table 7-4

	Proton	Neutron	Difference in Mass
Mass	1.67262E-27	1.67493E-27	2.3055E-30
	6+ orbiters	3+orbiters	
N/6 Poisson =	1.75391E+20	4.88657E+20	=n/3 Poisson
Alpha	0.9999999999563	0.9999999999476	
N Radius	1.85409E+11	1.54739E+11	meters
Radius	8.76801E-16	1.15E-15	a Ln(a)/sqrt(N+6)
V^2 = R gu	3.27547E+15	3.47891E+15	(GM+Kq)P/R
Velocity in orbit	57231712.77	58982305.63	M/sec
1/(1-v/c)^2)	1.037822946	1.040266772	Relativistic Mass Fac
Rest Mp/1836.15/6	1.51823E-31	3.04065E-31	Rest Mn/1836.15/3
Average	2.27944E-31	2.27944E-31	rest mass
Single orbiter rest	2.46637E-31	2.46637E-31	mass, Calculated

What is of great interest is that we have employed the parametric equations used in chapter two to indicate the radii from the masses in Mev of electrons and some pions. The masses and radii of the proton, neutron, are known. The method uses central masses that are depressed for the orbiters (not shown) although the small mass of the orbiting charges is insignificant in the computation.

Statistical mechanics emanated from trying to understand *the one way* thermodynamic motion of heat from hot to cold and the two way nature or reversal of the arrow of time inherent in Newton's laws of motion. Ludwig Boltzman developed entropy as a probabilistic state of an ensemble. Some say this was negated by the fact that a state could revert to an original state of low entropy from high entropy given enough time and noted that a pack of shuffled cards might just do that if reshuffled often enough. However, entropy is now defined:

$$S = K_B \ln\left(\frac{1}{V}\right) \qquad\qquad 7\text{-}9$$

It would be nice if equation 7-9 were exacting. Two things mitigate between entropy and change in entropy. First and perhaps foremost is that the atoms on both sides of an equation are balanced. So much for change in entropy because it would be nil. Second, the apparent Boltzman constant is not always that used in the literature. That is, different values of entropy for the same compounds are seen to exist in the literature. What is needed is an assessment of the "entropy contributions" of various bond types. This is a subject that is to be pursued in writing a textbook in physical chemistry. A possible approach is to use the reciprocal densities to obtain the change in volumes of any reaction which could possibly yield the correct entropy change. Hopefully Boltzman's constant will apply. It would be possible then to define reciprocal densities can be used to obtain volumes. Temperature effects have been discussed elsewhere in the book so that the reciprocal densities at various temperatures can be used to assess entropy at these points. I hope that someone tries this and lets me know before I get around to it!

The orbital velocities that were higher than expected due to dark matte can be contrasted to the normal velocities that we would expect due to normal expectation of Newton's law are depicted in Figure 7-1. Since the orbiting mass is nominally unity the extra mass, if real has to be at the central mass. If the anomaly is not due to the extra electrons on the orbiting mass causing a greater pull by the protons of the central mass then there may be an unobserved black hole that is the culprit. The alternative of a new kind of matter is not of consequence because this would entail revision of the creation of atoms in the stars.

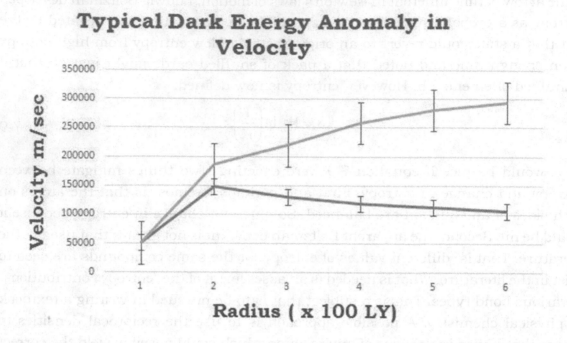

Typical Dark Energy Anomaly in Velocity

My view of cosmology falls short of the usual presentations. For one I believe that using MKS units is fine. When one encounters the units that are used to supposedly keep the numbers small I counter with the idea that the powers of ten were designed to do just that. I cannot find much data on the internet regarding redshift of specific stars. Why is this?

The application of Newton's law that resulted in the Friedman equation, equation 7-10, was arrived at by using a Lagrangian. Presumably this leads to a wave. I prefer the equation in the following form without the extra term that Friedman use which a constant was due to the use of a Lagrangian.

$$H^2 = \frac{4\pi}{3} G\rho_m + \text{constant} \qquad \text{7-10}$$

The Hubble constant, H^2, is the square of velocity over the square of radius. Equation 7-10 each term of which has the units of the Hubble constant has no vector directed outward; nor, is it constant over two volumes if density is not. We do not know what to expect for the Poisson if it is not a direct multiplier of G because if it is involved with the Lagrangian and is in the trailer constant then it is misplaced. We note however that the density contains a multiple the mass. Should the distance from the source grow so great that the velocity approaches the speed of light we might speculate that the velocity is really on such a large circle about the emitter that it may be hypothetical and/or virtual. This is a problem at high redshift. At this point as the light reaches earth it is struggling against the g field of the ever increasing mass that is in this circle even if it is blued somewhat as it enters earth space. The Doppler principle of light is somewhat misunderstood.

When the Doppler shift of sound occurs there is a great shift in the sound velocity that reaches the ear during an extremely short time span. We also call that phenomenon deceleration. The photon is doing a gradual shift against an ever increasing amount of central mass as it travels to earth. We lose frequency as deceleration occurs in both with sound and light. A slight blue shift ameliorates the photon energy loss as it approaches earth. The frequencies in light and sound both entail acceleration or deceleration. As radius from the source of light increases the increasing central mass causes the photon to lose frequency, working against an increasing g. Neither source nor the observer of light are essentially in motion. When a constant motion singular frequency source of sound approaches the pitch or frequency increases slightly because of the motion of the source. When the sound and source pass there is a reversal of the velocity of sound a sudden great deceleration. This results in a large Doppler shift which is a sudden reduction of the pulses of frequency. Thereafter as the source of sound recedes a continually lesser reduction of frequency is heard. Light is generally ever approaching frequency is not due to recession because as the electron moves up and away from the source it moves against g. Velocity of the photon does not change as a ball does when thrown upward against gravity and loses energy. Instead it loses energy by loss of frequency that is, gains redshift, as it works against g. The greatest change in frequency of sound is nearly instantaneous at the instant the incoming sound passes us suffering nearly instantaneous deceleration. Thus the Doppler principle of sound and light do not have an exact analogy.

Chapter Eight

Recombination of Helium Ions in a Spark Chamber

In 1966 this study was done with a spark chamber that was built by Dr. James B. Westgard of the Physics Department of Indiana State University. Dr. Harold Hughes, a student of Rabi at Columbia, was Department Chairman at the time. The basis for the study of the rate of the recombination due to return of ionized electrons with their rare gas ions was initiated by Keny in 1928.[67] Using the Langmuir positive probe technique at 800 microns pressure he found a value of 2×10^{-1} cm^3/ion-sec for the rate of his variable. The data of Biondi and Brown for helium ions gives a value of 1.7 times 10^{-8}.[68] The initial value of concentration was determined by its proportionality to the square rood of the intensity of emitted light. Three independent equations confirmed this order of magnitude, one a spectrophotometric technique; two, a combined probe and spectrophotometric method; and finally, an alternate calculation by the second method.

My study utilized the spark chamber built by Dr. Westgaard who was in charge of the laboratory and who taught a course in electromagnetic theory. The concentration was determined by the capacitive voltage drop ration per unit charge per cm, divided by the initial cross section for diffusion at zero time. The cross section is determined by classical kinetic theory applied to the rate of the Reynolds to the Schmidt number.

An attempt to measure the recombination of argon ions with their electrons using a Teflon interlayer between steel plates with a .34 cm gap containing 19.2 ml of argon gave a signal of such low magnitude that if interpreted versus α would comprise some 2 % of the some 15 volt signal taken from the capacitor of the differentiated resister voltage from the steel plates to ground. An expression known to Townsend is known to correlate the variation of recombination coefficient with pressure:[69]

---

[67] Loeb,L. B., Fundamental Process of electrical Discharge in Gases, New York Wiley, © 1939, p. 150

[68] Biondi, Max and S.C. Brown Phys Rev. 76,11,1647 1949

[69] Korff, S.K. and M. Kallman, Electron and Nuclear Counters, Princeton, N.J. D. Van Nostrand Co., Inc., p. 5

$$\alpha = zpe^{-Wp/E} \qquad \text{8-1}$$

This empirical equation requires an external field effect present in early recombination and mobility studies which, due to present technique, is only intermittently applied in order to sample the concentrations. The avalanche factor Z of this equation can be estimated by the loss of voltage on sparking. The spark chamber was operated in a voltage region where 10^{13} ions/cm^3 were obtained and an average over the voltage region was taken so that an average avalanche factor accrued which cancels in treating the data. The voltage region was less than 4000 volts above the threshold voltage. This condition gave the same order of magnitude of concentration that was created by x-ray or magnetron bursts.[70] Since these initial concentrations essentially determine the actual value of α quoted in this report, the excellent agreement with literature values may point to the existence of an upper limit on the number of ions a spark can initially sustain. The delay time after the spark is initiated until a concentration is sampled appears to be shorter than for any other data found in the literature.

Two books, one by Loeb and another by Darrow, both are a good account of the theory up to 1937.[71] A series by Compton and Langmuir is of interest since the competing equations of Thomson and Langevin are presented.[72] These equations contain corrections to what is a second order reaction rate of kinetic theory.

Diffusion is related to mobility through the equation below where e is to the charge on the electron and δ is the mobility or v/E.[73]

$$\frac{\delta}{D} = \frac{e}{KT} \qquad \text{8-2}$$

The knowledge of diffusion required in this study is found from the differential equation of Sherwood.[74] Integration of $-c\ dx/dt = K\ c\ dx$, where K is proportional to α leads value for the diffusion cross section and from the Reynolds to Schmidt number ratio, to the diffusivity, D. The formula of Meyer allows a comparison of diffusivities to be made.[75]

$$D_{12} = \frac{N1/\lambda1 + N2/\lambda2}{2(N1+N2)} \qquad \text{8-3}$$

[70] Loeb, L. S., op. cit., p. 13

[71] Darrow, K. K., Electrical Phenomena in Gases, Williams and Wilkins, p. 200-250,1932

[72] Compton, K. T., and I. Langmuir, Rev. Mod. Phys,, 191, 1930

[73] Thomson J. J., Conductivity of Electricity Through Gases 3rd ed., p 79

[74] Sherwood, T. K. and C.E. Reed, Applied Mathematics in Chemical Engineering, New York, McGraw-Hill, 1939, p. 47

[75] Jeans, J., Kinetic Theory of Gases, Cambridge 1967, p. 201

Several types of recombination are suggested by Loeb.[76] These differ in the equations employed and the type of distribution. He lists initial, preferential, coplanar, and volume recombination's as being representative kinds. Jeans notes that relaxation times of nanoseconds in air ensure that initial, random, or Gaussian distributions are, in several relaxation times, essentially converted to Maxwellian distributions.[77]

Preferential recombination or recombination of the same elemental particles, as an idea, arose from the effect later found by Moulin to be due to the spatial network like distribution given by alpha particles.[78] Jaffe derived an equation for this columnar recombination and his equation when applied to the data of Schemel showed greater constancy with time of an observed to a calculated α when longer bursts of ionizing radiation were used.[79] His assumption of a Gaussian distribution with average displacement in conjunction with diffusion correction factors requires several experimentally determined constants.[80]

The mechanism of volume recombination is that of second order kinetics. A plot of 1/e versus time should give a slope proportional to α. Plimpton, Rümelin, Marshall, Luhr, Gardiner, and Sayers all found higher initial values for α.[81] Since mobilities of electrons are 10^6 greater than those of ions, the assumption that diffusing ions (electrons) fill the available volume in the case of rare gases would be tenable but for the positive ions and the necessity for recombination in such short times. Again mobility, classically, is in terms of the average drift under a field which of course would alter the distribution if present. Wahlin measured the limiting mobility of electrons as a field was approaching the limiting mobility of elections at zero field in helium and reports the value 1.086×10^4 cm/sec.[82]

Beta and gamma rays give 10^2 to 10^3 ion pairs per cm and at 60 KV. There are 200 ion pairs per cm length, representing a concentration of 3×10^8 to 3×10^6 ions per cm.[83] Electron recombination gives rise to the correlation of $\Delta E = hf$ whereby an electron goes from the short wavelength limit of a series as it falls from infinity into orbit.[84] A continuous spectrum ensures if the electron has initial momentum. This was shown by Bartels to be a preferential or self-recombination phenomenon

---

[76] Loeb, L. B., op. cit., p 132-146

[77] Jeans, L., op. cit., p 103ff

[78] Moulin, Ann, Chim. Phys., 21550, 199910; 22, 26, 1911

[79] Loeb, L. B., loc. cit.,

[80] ibid., p. 139

[81] Hendren, L. L., Phys. Rev., 21, 344, 1905

[82] Darrow, K. K., op. cit., p. 232

[83] ibid., p. 131-132

[84] ibid., p. 149

in certain series of alkaline earths since it was responsible for certain parts of the continuum.[85]

A source of beta or x-rays changes the initial pattern of distribution giving the electron the high velocity needed for continuous radiation beyond the $E\infty$, and increasing the probability of electrons attachment to form a complex ion, thus disrupting the statistically uniform distribution inside a sphere of attraction shielded by an outer cone of attached ions.[86]

A series of reactions for argon is discussed by Bond.[87] In one reversible reaction possible, a mechanism involving an Ar^+ion is postulated as possibly being important, though possibly present in negligibly small amounts. Frost and Pearson have summarized an equation useful for parallel first and second order reactions which this is.[88] Since the first rate is controlling in the present study it becomes unequal and consecutive of an unequal concentration form.

Korf uses the following relation where C is the capacitance in farads, to describe the number of ions collected.[89] The ratio of electron mean free path to the ionic mean free path is $4\sqrt{2}$.[90]

$$dV = \frac{dq}{dc} = 1.6021 \times 10^{-19} c / C \qquad\qquad 8\text{-}4$$

The purpose of the study was to investigate the nature of spark discharge in helium and argon in order to design a liquid filled container suitable for operation in the proportional region as a counter in a variety of measurement situations. Accordingly a measure of the recombination time was sought and argon being the more dense rare gas than air was selected, for calibration purposes, to test a proposed chamber design. An electronic circuit was devised by Dr. Westgaard. It is shown photographically as Figure 81A.

For Helium

The modified circuitry is shown in Figure 8-3 and a list of equipment is given at the end of the chapter. Helium was allowed to pass through four levels of the purged spark chamber at about 5 to 10 ml/min. One of the gate generators allowed only one pulse per second to activate the pulser. This can be seen as Figures 8-1B, 8-2 and at the end of the chapter. Many of these coincidences did not cause any level of

---

[85] Bartels, S., Physik, 105, 74, 1937

[86] Loeb, L. B. op. cit., p. 132

[87] Bond, J. W., Phys. Rev., 105, 1683, 1957

[88] Prout, A. A., and R. G. Pearson, Kinetics and Mechanisms, New York, Wiley, 1953, p. 165

[89] Korff, S.A., op. cit., p. 8

[90] ibid., p. 37

sparking and some of those which did were not counted as cosmic inspired pulses even though they were definitely counted as coincidences.

Spark Criterion

The criterion for a cosmic inspired spark was that it encompass at least two section heights of the chamber with reasonable linearity and virtually no paralleling outside its sometime slanted path. The counter registered the number of coincidences. An eye count of observed sparks conforming to the above criterion was recorded in ratio to the number of coincidences shown by the counter during this time.

Raw Data

As the raw data recorded in Table 1 show, an eye count of 25 such cosmic sparks conforming to the above criterion and the number of coincidences recorded as a function of the delay from event initiation at voltages between 3 and 7 KV above the chamber threshold. The coincidences increase with delay, therefore, as expected after 1 microsecond, the entire data, which is presented further was taken up to 100 microseconds. Table II gives the raw data converted to eye counts per 100 coincidences.[36] A statistical breakdown of this data indicated the wisdom of dropping those counts at the lowest voltage of 3.5 KV. A glance at the data indicated that the threshold voltage is diagonal through the data which gives some doubt as to the terminal data point.

Table 8-1 Eye Counts per Coincidence vs Delay

Volt/KV		1	10	20	30	40	50	60	70	80	90	100
6.0		25/44	25/40*	25/46	25/46	25/47	25/53	25/68	12/48	5/40	5/56	3/37
5.5	*	25/45	25/39	25/44	25/48	25/45	25/42	25/73	12/51	5/46		
5.0		25/43	25/45	25/38	25/39	25/50	25/48	25/59	13/42	5/51		
4.5	***	25/41	25/48	25/48	25/50	25/48	25/59	25/97	5/47	5/46		
4.0		25 46/25/39	25/48	25/54	25/55	25/41	25/55	25/50				
3.5	Not Avg		25/62	25/46	25/74	25/60	12/27	25/61	10/109	12/92		
*	@7 Kv											
**	Av of 3.5 4.5											
***	At 6.5 KV											
3.0	No Spark											
AVG	25/per		45.8	46	52	50	52.5	86.1	193	215	294.2	

The base counts were not taken at the same voltages as the remainder of the data in two of five instances but the average of voltages is supposedly centered about a mean at a given delay time and the count at one of the high out voltages was below the mean of the five lower voltages in the sample. A Cogito calculator was used to calculate all values associated with the known up to 12 decimal places. The least accurate measure was that of the voltmeter.

Figures 8-1 to 8-7 either follow or are in the appendix. The cage about the reaction chamber was needed for safety because 10,000 volts jump easily and it was necessary to be close to the reaction chamber. Photographs are available in Figures 88-1 a including 8-1 b of the chamber and the cage.

Procedure and Circuitry

A cylinder of argon was allowed to purge a 10 x 15 x 4 inch plastic b ox covered with a Formica top having a 5 x 10 inch window. The circuit is shown in Figure 8-3. After 10 minutes at 1.5 to 2 liters/min the chocks from the top plate of the two 3.178 x 3.97 inch plates were pulled and a 41 mmf capacitor filled with 19.2 ml of argon was thus in series with a variable 50 to 5K resistor to ground. The action of the circuit was to place a charge on capacitor C_1 which discharged to ground through variable R_2 of from 8 to 1 mega ohms charging the argon capacitor to negative potential which was equal substantially to the from zero to 10 KV available from the power supply. The time lapse from initiation to the placing of voltage across the plates by action of a pair of coincidence counters was negligibly small. Resistances were varied through the". The cage about the reaction chamber was required because in order to take data the reactor it was necessary to be close and the voltage used might otherwise jump to the viewer. At the end of the chapter are schematics of Figures 8-1 a, b. There is a Figure 8-5, List of Equipment and circuitry in Figures 8-6 and 8-7. When it became apparent that the .34 cm spark gap of the Argon chamber was too short and the signal was being lost in the carrier voltage, essentially insofar as accurate measurement combination of ranges indicated. On the oscilloscope the voltage drop characteristically was highly variable in its appearance with the spark tube action; and, while its magnitude was statistically variable with time and the fastest resolution sweep of .1 microseconds gave breakdown at variable slope heights, no eye discernable pattern of reduction was found as the delay time of sampling was varied from .2 microseconds to 1 second. It was calculated that a 2 % variation superimposed on an already statistical distribution of that or greater would have to have been found to achieve the literature value of Kitne.[91]

---

[91] British Chemical Engineering Volume 6, p 742

Table 8-2 Coincidences per 100 Eye Counts

V/Kv	1	10	20	30	40	50	60	70	80	90	100
6.5	176	160	184	184	188	212	178	400	800	1120	2900
5.5	180	156	176	192	186	168	292	425	920	1210	
5.0	172	152	156	200	192	236	315	1020			
4.5	164	192	188	156	236	200	288	940			
4.7	184	192	216	220	164	220	284	1090			
Avg	175.2	170.4	184	190.4	207.2	210.2	310	775	860	1100	2900

Table 8-3 A First and Second Order Rate Data

Run	Elapsed Time	c/c_0	Kt	$K1 \times 10^{-3}$	$1/c$	$K2$	
1	10	0.9556	0.0450	5.0433	1.0464		
2	20	0.9522	0.0490	2.5798	1.0502		
3	30	0.9202	0.0832	2.8699	1.0858		
4	40	0.9125	0.0916	2.3480	1.0959		
5	50	0.8456	0.1678	3.4238	1.1826	0.83171	1-5
6	60	0.5648	0.5713	9.6829	1.7705	11.2373	1-6
7	70	0.2261	1.4869	21.5498	4.4233		
8	80	0.2037	1.5910	20.1394	4.9087		
9	90	0.1510	1.8802	21.1263	6.6210	61.358	7-9
10	100	0.0604	2.8065	29.3489	16.5525		6-10

Multiple Reaction Rate Theory

The following reaction is postulated in which the first reaction is pseudo first order due to the large concentration of helium. Initially at least, the second reaction is second order and of equal concentration for the purpose of integrating the rate equation:[27]

$$\mathrm{He_2^+ \rightarrow He_2^+ + 2e^- \rightarrow 2\,He}$$

8-5

$$-\frac{dx}{dt} = k_1(A-x) + k_2(A-x)^2$$

8-6

Assume $k_1 / k_2 = r$, and then by Dwight Integral 160.01 and by the application of the boundary condition that B-0 and A=1 when $k_1 = k_2$:

$$-k_1 t = \frac{1}{r+(B-A)} \mathrm{Ln} \left[\frac{A}{(B+1/r)} \otimes \frac{B+(1/(r-x))}{A-x} \right]$$

8-7

By application of a second boundary condition to the foregoing equation, namely, that when the above equation is correct and defined in terms of either A or B and by applying limits and noting that the first factor is unity we have:

$$-kt = \text{Ln}\left[\frac{A(r-x)}{r(A-x)}\right] \quad \text{t<59 ; x<r ie B=0} \qquad \text{8-8}$$

$$kt = \text{Ln}\left[\frac{Ar}{(x+r)(A-x)}\right] \quad \text{t>59; x>r; B} \approx \text{x} \qquad \text{8-9}$$

Time is in microseconds for the quoted interval. The equations give for A-1 and r-1:

$$(1-x) = e^{-\overline{K}_1 t} \quad \overline{K} = k/2 \qquad \text{8-10}$$

$$\frac{(1-x)}{(1-x^2)} = e^{-\overline{K}_2} \quad \text{and} \quad \overline{K}_1 = \overline{K}_2 \qquad \text{8-11}$$

Where $\overline{K}$ is the average or least mean square slope divided by two. The above equations are graphed in Figure 8-4 versus the data for $\overline{K}$ of 3.656 the average as shown, including the leas squares value of 3.848 x 10^{-3} which is equal in magnitude to a first order first reaction and a second order second reaction by the above assumption.[48] The actual second order least squares k_2 has value 1.8238 x 10^5 and the ratio $\overline{K}/k_2$ .02111 and will be shown to be useful later.

Diffusion of Helium Ions

Sherwood gives the relation $-d^2 c / dx^2 = kc / D$ for simultaneous chemical reaction and diffusion where if $k = \overline{K}$ the equation may be integrated since $\overline{K}$ is pseudo first order and constant giving:[8]

$$c / c_o = e^{-kx^2/D} \qquad \text{8-12}$$

And, since k is linear in x, then

$$c / c_o = e^{-\overline{K}x^2/D} \qquad \text{8-13}$$

In order to evaluate x^2 we want to have a measure of D and this is provided in the ratio of the Reynolds to Schmidt numbers.

Postulate:
$$\frac{\lambda u \gamma^2}{DN^2} = \text{Ln}(c/c_o) \qquad \text{8-14}$$

The left hand side is the ration of the Reynolds to Schmidt numbers. The value of $\frac{\gamma^2}{N^2} = 16.8^2/19.4^2$ (or 0.7499) for helium at 292°K. The velocity, u, of He$_2^+$ion, which is found to be more appropriate is its relative r.ms velocity to that of its electron.

$$u \approx 1.224\sqrt{2r/Mg} \qquad\qquad 8\text{-}15$$

If $u_e = u\left[e^{T_e T_{He}} - 1\right]$ then for $T_e T_{He} = z = 1$ we have $u_r = u(e^z = 1)$ so that $u_r = 1.226 RT/g\left[e^z - 1\right]$ and $1.224\sqrt{0.315(292x10^7/980}(1.71828) = 10.46852x10^3$ which leaves for u the value 6.09244 x 10^3 cm/sec (r.m.s.).

Now the mean free path of the electron based on a hybrid radius of 74 Å, since it is $4\sqrt{2}$ times that of the positive nucleus is:[37]

$$\lambda_o = \frac{4(2.32T \times 10^{-4})\sqrt{2}}{\sigma_h^2} \qquad\qquad 8\text{-}16$$

$$= \frac{9.288(292)\sqrt{2} \times 10^{-4}}{.74^2 \times (760}$$

$$= 1.666 \times 10^{-3} \text{ cm}$$

If the hydrogen radius were used and a mean free path of particle is chosen, one obtains and expression only 3 % less. This is a minimum radius for maximum mean free path. Use of the Wahlin velocity of 10.84 x 10^{-3} gives a 3 % greater λ (and thus α as well than the λ_e above. In Figure 8-3 the data from 10 to 60 is a plot of (1-x) = exp (3560 t

Second Order Fit to Pseudo First Order Reaction

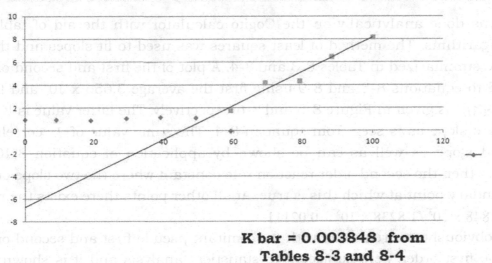

K bar = 0.003848 from
Tables 8-3 and 8-4

Figure 8-2

Disposition of the Table 11 Data

The averages of coincidences of Table 11 were subjected to tests for first and second order kinetic rate constants:

$$c/c_o = e^{-Kt} \quad \text{(First order)} \qquad 8\text{-}17$$

$$\frac{1}{c} = \frac{1}{c_o} = Kt \quad \text{(Second order)} \qquad 8\text{-}18$$

Table 8-4 Average Recombination Coefficients

	Runs	K x 10⁻³ First Order	σ		K₂ ***
First Order	1-5	3.293	0.965		0.83172
First	1-6	4.324	2.35	**	11.2373
Reaction	2-5	2.805	0.402		
Least Square		3.657			
Arrhenius equation		3.848	Meas'd Value	= 7.314	
Second Order	6-10	20.189	5.995		182.38
Second		18.123	4.900		
		20.939	0.5909		61.358
**Run 6 was outside					
2 sigma with1-6					

This was done analytically on the Cogito calculator with the aid of table of natural logarithms. The method of least squares was used to fit slopes and these results are summarized in Tables 8-3 and 8-4. A plot of the first and second order data fitted to equations 8-8 and 8-9 using first the average 3.656×10^3 and then $\overline{K} = 3.848 \times 10^3$ is given in Figure 8-5 and 8-6 respectively. The latter value is ½ the least square slope as is seen from equation 8-4. The same value of $\overline{K}$ correlates the second slope as well as can be shown by application of equation 8-10. If $\overline{K} = \overline{K_1} = \overline{K_2}$ then the second order reaction rate constant where the two slope cross is the boundary point at which this is true. At all other points there exists the ratio $\overline{K}/K_2 = 3.848 \times 10^3 / 1.8238 \times 10^5 = 0.02111$.

There obviously two reactions with concomitant pseudo first and second order slopes, The first order were subjected to statistical analysis and it is shown the sixth data point belongs mostly to the second data group using a 2σ criterion. This is shown in Table 8-5.[41]

It is the pseudo first order rate constant from the first reaction which is useful in evaluating the value of x^2 of equation 8-10 and allows us to put K_2 on an ions per cm³ basis which is the requisite α. In this connection the aforementioned ratio of $\overline{K}$ to K_2 occurs and is found theoretically by application of the Arrhenius equation since the ratio so obtained is so close to the least mean square of $\overline{K}$ from equation 8-9 and the average $\overline{K}$ of runs one to five.[38] The task is to find the initial cross section since ions per spark length are given by theory and the spark chamber constants which influence the voltage drop on sparking.

Initial Cross Section from Equation 8-10

$$\eth x^3 = (\eth \lambda_e u \gamma^2)\sqrt{\overline{K}}N^2$$

$$= \frac{\pi(1.666 \times 10^{-3}\ 6.0924 \times 10^3\ 0.7499)\,x10^{-3}}{3.848}$$

$$= 6.2169 \times 10^{-3}$$

8-19

Since $c' = d\text{V C/eM}$ ions per cm where $C = e_o A/L$ farads and $\alpha = \pi \overline{K} x^2 / c'$ ions per cm and then at the boundary:

$$\alpha = \pi K_2 \lambda_e u_r \gamma_2 LeMN^2 e_o A \sqrt{K}\ dV$$

8-20

Equation 8-20 applies at the boundary and at all other points as well since by the Arrhenius equation:

$$\frac{\overline{K}}{K_2} = e^{\frac{-\lambda_2}{(\lambda_1 - \lambda_2)}}$$

8-21

With the wavelengths λ_1 and λ_2 equal to 1215 and 1640 Å respectively. Equation 8-21 yields the value 0.02111 which allows an estimate of α by combining equations 8-19 and 8-20 since C=47.84 mmf and $dV e_o A/LeM = 1.2105 \times 10^{11}$ then:

$$\alpha = \frac{\pi(1.666 \times 10^{-3}(6.092 \times 10^3)(1.7182)(1.602 \times 10^{-19})(0.7499))}{(0.02111)(47.837 \times 10^{-2})(3000(1.2105 \times 10^{11}))}$$

$$= 1.329 \times 10^{-8}$$

This is theoretically correct for a hybrid 0.74 Å radius and an electron mean free path.

Discussion of Optimal Electron Mean Free Path

Using the velocity of Wahlin and without the reduced mass concept the value 1.34×10^{-8} for α was found. Decreasing the radius of the hybrid and obtaining the same mean free path requires using the particle mean free path, which is $4\sqrt{2}$ less giving a value of the recombination constant less by this factor. If, for example, the hydrogen radius were employed there is no classical factor which yields the literature result and whose value is based on what can be called a mean free path. Similarly, if the helium radius is employed the lowering of α thereby has no factor

141

larger than is being classically employed in the theoretical value quoted to cover the reduction.

Inter-Diffusivity Electrons & He+ ions in He at 1 Atm and 292 °K

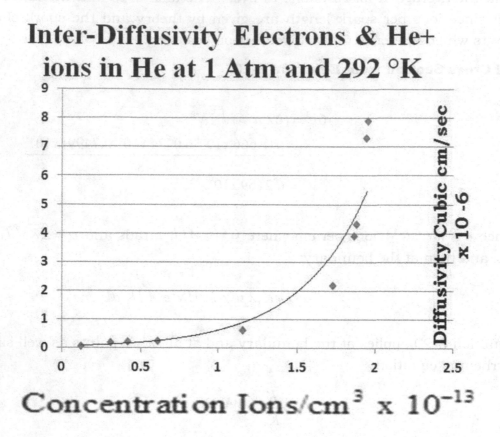

Results of the Measurement of Recombination Coefficient

By noting the effect of the experimental differences in averages for $\overline{K}$ an amount equal to 10.7 x 10^{-3}, its value is based on u_r, an analysis of the agreement with the value of Biondi and Brown is easily made.[43] The method hinges on the correctness of the measured slope of K_2 by the least square method and the goodness of fit to the pseudo first order initial reaction. This involves also, the voltage measurement, the spark chamber constants, the validity of the cross section found, the kinetic concepts of hybridized mean free path of the electron assumptions, the relative velocity concept, the temperature and pressure determined constants going into the Reynolds and Schmidt numbers. All these enter into the result. The results are in Table 8-7.[44] When one subjects the 3.253 x 10^3 value of rounds 1 to 5, the closest experimental agreement to the literature value becomes 1.64 x 10^{-8} cm^3 / ion sec. These values are of course; subject to the ± 10 % of voltage. There are three supporting facts to substantiate the foregoing results. First, the diffusivity calculated should and does agree with the literature values. Second, an independent check of the diffusivity using Meyer's equation for diffusion substantiates the first result. Finally, an energy transition for helium, helium ion interaction tends to account for the mechanism and suggests that the diffusing particle moves in the center of an eight atom complex with a 29 Å neighboring atom distance and having an average bond comparable to a halogen to cation bond in strength.

Diffusion

Since by equations 8-10 $c/c_o = e^{-\overline{K}t}$ and postulating from equation 8-10 that $Ln(c/c_o) - \overline{K}x^2/D$ we have:

$$D = \frac{-\overline{K}x^2}{D} \qquad\qquad 8\text{-}22$$

(Kt) is the experimental value. Again, postulating from equation 8-10, since $x^2 = \lambda_e u_r \gamma^2 / N^2$ where the ratio of K values is so as to assure low diffusivities at low concentrations, therefore:

$$D = \frac{\overline{K}\lambda_e u_r \gamma^2}{N^2} \qquad\qquad 8\text{-}23$$

Table 8-5 Inter-diffusivity of Electrons and Helium ions in Helium at 1 Atmosphere & 292 °K

Time x 10^6	Ions/cm³ x 10^{-13}	K x 10^2	D cm²/sec x 10^{-6}	
1	2.0375			
10	1.9473			
20	1.9400			
30	1.9742			
40	1.8592			
50	1.7288			
60	1.1500	57.120	0.628	
	(0.86)		0.418	Biondi & Brown[41]
70	0.6046	148.694	0.241	
	(0.57)		0.235	Formula of Meyer[9]
80	0.4151	159.101	0.225	
90	0.3077	188.024	0.191	
100	0.1321	289655	0.059	

$$D = \frac{3.56 \times 10^3 \, x1.666 \times 10^{-3} \; 6.0923 \, x \, 0.7499 \times 10^3}{Kt} \qquad\qquad 8\text{-}24$$

Values of diffusivity from the formula of 8-4 are listed in Table 8-5 and are plotted in Figure 8-6 above. The diffusivities given by the formula of Meyer are shown as two points on the curve, the value D from the mean free path of the hybrid radius occurring and time of onset of the second reactions, while the value of D from the mean free path determined from the helium radius is at the foot of the curve. Reduced mass velocities were used for the first; and, velocities which were the $\sqrt{2}$ less were employed in the second with ratios of $(e^z - 1)$ were used for the velocities of helium, helium ion, and electron.[9]

$$D_{12} = u_e \lambda_e + u\lambda \qquad\qquad\qquad \text{8-25}$$

$$D_{12} = (16.556 \times 10^3 \ x \ 1.667 \times 10^{-3} + 6.0294 \times 10^3 \ x \ 0.2946 \times 10^{-3})24000 \qquad \text{8-26}$$

$$= 0.235 \ cm^3 / \sec \quad (x10^6)^* \qquad\qquad\qquad \text{8-27}$$

And for helium radius where electrons are diffusing in helium:

$$= (16.56 \times 10^3 \ x \ 4.167 \times 10^{-4} \ x + 6.092 \times 10^3 \ x \ 0.736 \times 10^{-3})2400 \qquad \text{8-28}$$
$$= 0.041 \ cm^3 / \sec \quad (x10^6) \qquad\qquad\qquad\qquad \text{8-29}$$

Biondi and Brown recorded a value of 10 mm of mercury which if raised to atmospheric pressure would be 0.418 * which is about twice the value given by the formula of Meyer for electron diffusing helium ions (hybridized radius) at a steep portion on the curve.[42, 46]

Finally Figure 8 is a diagram of the energy level for the two postulated mechanisms:

$$He^+ + He \rightarrow He_2^+ \quad + 2.68 \ ev \qquad\qquad \text{8-30}$$
$$He_2^+ + e^- \rightarrow 2 \ He \quad -0.345 \ ev \qquad\qquad \text{8-31}$$

The transition in equation 8-30 is between n = ∞ and n = 2 of the series:

$$\frac{10^5}{1.215} = \left[\frac{1.0967756 \times 10^5}{e^{e/10}} \right] \left[\frac{1}{1^2 - \frac{1}{n^2}} \right] \qquad\qquad \text{8-32}$$

The above equation is satisfied if n equals ∞. The initial member is the 1640 Å line so that the first reaction would be an increasing orbit for the helium ion, which, assuming it has the hydrogen radius with one electron, would give an intermediate value. The helium could undergo a 5 to 2 Paschen transition for which the required energy would need eight participating atoms to render a null energy effect. The 1215 Å line is also the initial member of the Lyman Series. The value of the factor in 8-32 is 0.74488166.

Discussion

The correlation found between the physical properties of helium and the recombination of helium and its first ionization product has been presented. The reaction takes place in two steps through a positively charged activated complex of the helium nuclei essentially within 10^4 seconds from its formation by radiation in an atmosphere of helium at 292 °F. Diffusion calculations confirm the result and the value found for the second step recombination coefficient is $1.6 \times 10^{-8} cm^3$/ion sec compared to the value 1.7×10^{-8} reported in 1959 by Biondi and Brown in a method wherein no data was taken until 10^{-3} seconds after an ionizing burst of radiation.

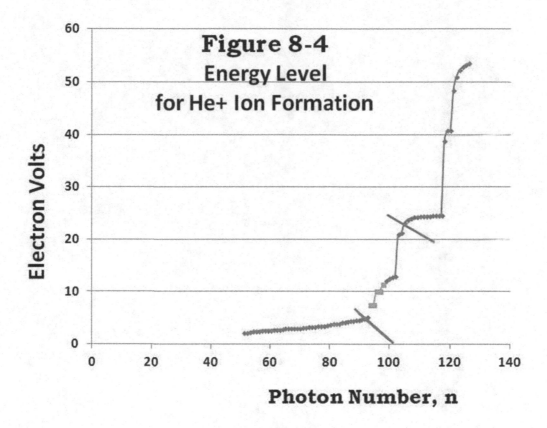

Figure 8-4
Energy Level
for He+ Ion Formation

A theoretical method of calculation based on knowledge of the physical properties, the wavelengths and the voltage drop on sparking, was developed using mostly classical concepts. The result, 1.4×10^{-4} times 13.5, varies mainly as the uncertainty in the voltmeter reading. Experimental results varied from 78 to 104 % of the literature values, depending on the method of averaging used. Diffusivity was $5.5 \times 10^5 cm^3$/sec and compares well with the value 4.2×10^5 of Biondi and Brown. A reaction of this ion type has been postulated for argon and the wavelengths involved are listed for helium.

Figure 8-4 indicates the small size of the region of operation. Ionization was occurring only in the area of the 94th to 105th photon. This treatment of the data was not found within the literature. The Reynolds to Schmidt number ratio coupled with the concept of activity should lend itself readily to liquid passé analysis.

Appendix

Physics for the Twenty-First Century

Reclassifying Some Iconoclasms of the Twentieth Century

An Introduction to Attribute Mechanics

J. Oswald Brooks
©2009

Physics for the 21st Century

Reclassifying Some Iconoclasms of the Twentieth Century

An Introduction to Attribute Mechanics

Jonathan Oswald Brooks

©2009

▶

Einstein Spent his Latter Years Searching for a Link Between Gravity and the Electromagnetic Field

There is a definite reason to believe that gravity is an electromagnetic phenomenon.

1. The Rydberg constant, as derived by Niels Bohr, can be rearranged; divided by the square of the unit electron mass, em² = 1²; and the units equated to G.

$$\frac{n\hbar V}{Mm} = G = \frac{KQq\sqrt{\dfrac{mVR}{h}}}{Mm}$$

Newton's Universal Constant is implicit in the Rydberg Equation

2. Equating the RHS to G and rearranging, noting that the radical has a ± unit jump value; and then dividing by R².

Local gravity is expressed in terms of Coulomb's Constant equally well.

$$g = \frac{GM}{R^2} = \frac{KQq}{R^2 m}$$

▶

An Imbalance of Charge Spewing from a Neutron Star Could Compromise Local Gravities

$$g = \frac{GM}{R^2} = \frac{KQq}{R^2\,m}$$

$$g = \frac{GM}{R^2\varepsilon_{rel}} = \frac{K\dfrac{Q}{M_{sun}}\dfrac{q}{m}M_{sun}\ \nabla(\phi)}{R^2}$$

$$G_{Solar} = K\frac{Q^+_{Sun}}{M_{Sun}}\frac{q^-_{Planet}}{m_{Planet}}$$

> **Einstein's General Relativity now can Host Coulomb's Law Constant**

$$G_{Extra\,Solar} = K\frac{Q_{Central}}{M_{Central}}\frac{q_{Entity}}{m_{Entity}}\ \nabla(\phi)$$

▶

Consider first the equality:

> Using calculable solar gravitational fields one can establish that the charge on the sun's protons requires a constant planetary charge to mass ratio.

▶ $$g_{solar} = \frac{KQ_{sun}}{R^2}\frac{q_{planet}}{m_{planet}}$$

$$g_{planet} = \frac{KQ_{planet}}{R^2}\frac{q_{of\,a\,kilogram}}{m_{one\,kilogram}}$$

> From the known mass of the planet the reduced fractional negative charge is calculable; and, this coupled with the positive charge on a kilogram of test mass, allows the known local gravities of the planet to be calculated.

▶

Data for Representative Planets

Planet g	Mass Kg	Q Planet	Q/Kg Planet	Radius to Sun	g about Sun	Q/Kg of Sun	Q Sun
Mercury 3.72	3.2 x E 17	2.49 E-11	7.75 E-29	5.79 E 10	3.98 E 02	9.58 E07	1.92 E38
Venus 8.92	4.89 E 18	3.79 E-10	7.75 E-29	1.08 E 11	1.14 E 02	9.58 E07	1.92 E38
Earth 9.8	5.97 E 18	4.63 E-10	7.75 E-29	1.50 E 11	5.96 E 03	9.58 E07	1.92 E38
Mars 3.72	6.45 E 17	5 E-11	7.75 E-29	2.28 E 11	2.57 E 03	9.58 E07	1.92 E38
Jupiter 24.8	1.89 E 21	1.47 E-07	7.75 E-29	7.78 E 11	2.21 E 04	9.58 E07	1.92 E38

Conclusion: G May Vary in Space

$$G = K \frac{Q^+}{M} \frac{q^-}{m}$$

$$GM = \{K \frac{Q^+}{M} \frac{q^-}{m}\} M$$

Note the use of the charge to mass ratio

1. This "new" collection of variables describing G may account for the "dark", presumably missing, matter in space due to local imbalances in charge compared to that of our solar system.

2. There thus may be only one force in nature; and, if so, it is electromagnetic.

Consider Next the Utility of the Negative Radical.

$$\frac{n\hbar V}{qMm} = G = \frac{\pm KQ^+ \sqrt{\frac{mvr}{\hbar}}}{Mm}$$

Not the first time the negative value of a radical has led to a breakthrough

$$\frac{GM + KQ^+}{R^2} = Unified\ Field$$

Note: It was necessary to employ charge to mass ratios in equating G, the excluded middle, to the electric field above.

Still two obstacles remain in order to allow evaluation of the Unified Field:

1. How to calculate atomic Radii

2. The assertion by P.A. M. Dirac that gravity is negligible

One circumferential orbit in time traverses 4 Radii

Since $\lambda = 4R$
There is never
an orbital mismatch.
The involved reciprocal
time of orbit is the
frequency !!

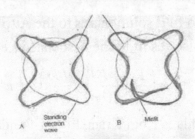

Conventional
orbit Mismatch

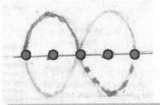

The $\left|\sqrt{\frac{mvr}{\hbar}}\right|$ is always unity

Conclusion: Never a Mismatch

Schrödinger's Wave Equation Initiated the New Era in Quantum Mechanics

Erwin Schrödinger

Erwin Schrödinger's radial wave equation was never solved for atomic radii until 1962 when it was transformed into a Bessel function in cylindrical coordinates by J. Oswald Brooks as the following slides will show.

$$\left(1-\varepsilon^2\right)\frac{d}{d\varepsilon}\left(\left(1-\varepsilon^2\right)\frac{dP}{d\varepsilon}\right) + l(l+1)\left(1-\varepsilon^2\right)P = m^2 P$$

Ziock, Klaus, Basic Quantum Mechanics, John Wiley and Sons, Inc.; New York, © 1969

One of the factors which delayed the solution was that the variable P (or ρ) that Schrödinger used for the radius had to be inverted in a function of another variable to give a viable radius. We will now show this in the process of achieving a *time average solution*.

▶

The Associated Legendre Equation is also found in another source.

To find solution sets to the Attribute Wave Equation we can also resort to the Associated Legendre Equation

$$J_n\left(k,r\right) = \left(\frac{1}{r}\right)\left(\frac{rd^2R}{dr} + \frac{dR}{dr}\right) + \left(k^2 - \left(\frac{n^2}{r^2}\right)R\right) = 0$$

This is from Kraut, Edgar, Fundamentals of Mathematical Physics, McGraw-Hill, New York,   1967, p 354 and Butkov, Eugene, Mathematical Physics, Addison-Wesley, Reading Massachusetts,   1968, p. 360

There are four solutions sets one of which will serve for atomic radii and another for radii of the solar system.

▶

152

Solutions of the d'Alembertian in Cylindrical Coordinates

The variable r of the d'Alembertian in cylindrical coordinates does not take the radial value; but, is the curvature. Replacing r with ρ :

$$J_n\left(k,\rho\right) = \left(\frac{1}{\rho}\right)\left(\frac{rd^2R}{d\rho} + \frac{dR}{d\rho}\right) + \left(k^2 - \left(\frac{n^2}{\rho^2}\right)R\right) = 0$$

The variable α can be expressed as an exponential:

$\alpha = e^x$ The solution set: $x = \{ k\rho, \frac{1}{k\rho}, \frac{\rho}{k}, \frac{k}{\rho} \}$

For an N of 100 the absolute value of the natural logarithm is almost halved; and at an N of 20 it is .142. Thus for a ten fold increase in k there must be a 20 fold increase in t. For the atom this eliminates all but $x = k/\rho$.

since we have $k = \sqrt{N+6}$ we now have $\alpha = e^{\frac{\sqrt{N+6}}{\rho}}$.

▶

Redefining the Associated Legendre Function

As now written, the equation below is given by Ziock to be the radial portion of Schrödinger's wave equation known as the associated Legendre equation, after substituting the constant A for $l(l+1)$:

$$\left(1-\varepsilon^2\right)\frac{d}{d\varepsilon}\left(\left(1-\varepsilon^2\right)\frac{dP}{d\varepsilon}\right) + B\left(1-\varepsilon^2\right)P = m^2P$$

We shall derive the Poisson by first letting the magnetic quantum number, m, be zero. Then we factor $(1-\varepsilon^2)$ out. Now substitute $\alpha = \varepsilon^2$, F for P, and we have:

$$BF + \frac{2d}{d\alpha}\left(2(1-\alpha)\frac{dF}{d\alpha}\right) = 0 \quad \text{since } d\alpha = 2\,d\varepsilon .$$

> **Note the substitution of α for ε^2 and F for P**

▶

Redefinition of Variables is Required to Achieve a Viable Equation.

$$BF + \frac{2d}{d\alpha}\left(2(1-\alpha)\frac{dF}{d\alpha}\right) = 0 \quad \text{Now let } A = \frac{B}{2}$$

$$AF + \frac{2d}{d\alpha}\left((1-\alpha)\frac{dF}{d\alpha}\right) = 0 \quad \text{Where } F = \frac{1}{1-\alpha}$$

$$-\frac{1}{c^2}\frac{\partial^2 \phi}{\partial t^2} = 0$$

$$AF + \frac{2d}{d\alpha}\left((1-\alpha)F'\right)$$

$$AF + \frac{2d}{d\alpha}\left((F)\right)$$

$AF + 2F'$ allowing A to equal N+5

$\nabla E = (N+5)\,F + 2\,F'$ we have the Poisson portion of the Attribute Wave

Note: equating of orthogonal Vector to zero

Orthogonal Vectors

F

F'

Zero Vector

Time average solution

The Scalar and Vector Functions are Known to be the Same for a Cylinder Function

Since $F = \dfrac{1}{1 - \dfrac{N-35}{N-28}}$ or $\dfrac{28-N}{7}$ and $F' = [\dfrac{28-N}{7}]$

Then $\nabla E = (N+5)\,F + 2\,F'$ or $\dfrac{(9N^2 - 271N + 588)}{49}$

In order to establish a solution to Ziock's equation

$$\left(1-\varepsilon^2\right)\frac{d}{d\varepsilon}\left(\left(1-\varepsilon^2\right)\frac{dP}{d\varepsilon}\right) + A\left(1-\varepsilon^2\right)P = m^2 P \text{ characterizing}$$

the associated Legendre function we start with the equivalent d'Alembertian in cylindrical coordinates:

$$J_n(k,r) = \left(\frac{1}{r}\right)\left(\frac{rd^2R}{dr} + \frac{dR}{dr}\right) + \left(k^2 - \left(\frac{n^2}{r^2}\right)R\right) = 0$$

Setting R = F, differentiating with respect to α, and setting n = r

$$J_n(k,r) = \left(\left(k^2 - 1\right)\right)F + F' = 0$$

results in N+5 $= k^2 - 1$ so $k = \sqrt{N+6}$

The Schrödinger Radius (Rho) is really the Curvature

The variable r of the d'Alembertian in cylindrical coordinates does not take the radial value; but, is the curvature. Replacing r with ρ :

$$J_n(k,\rho) = \left(\frac{1}{\rho}\right)\left(\frac{rd^2R}{d\rho} + \frac{dR}{d\rho}\right) + \left(k^2 - \left(\frac{n^2}{\rho^2}\right)R\right) = 0$$

The variable α can be expressed as an exponential:

$\alpha = e^x$ The solution set: x = $\{ k\rho, \frac{1}{k\rho}, \frac{\rho}{k}, \frac{k}{\rho} \}$

For an N of 100 the absolute value of the natural logarithm is almost halved; and at an N of 20 it is .142. Thus for a ten fold increase in k there must be a 20 fold increase in t. This eliminates all but x = k/ρ.

since we have k = $\sqrt{N+6}$ we now have $\alpha = e^{\frac{\sqrt{N+6}}{\rho}}$.

Inversion of Curvature in Circle of Invergence, α

Taking natural logarithms of $\alpha = e^{\frac{\sqrt{N+6}}{\rho}}$ we find:

$\rho = \dfrac{\sqrt{N+6}}{\alpha}$. Howard Eves book Survey of Geometry, vol.1, Allyn and Bacon, Boston, © 1963, p 145 ff notes that potential solutions often must be inverted. Therefore, the radial solution

is $R = -\dfrac{10\alpha\ln(\alpha)}{\sqrt{N+6}} \mathring{A}$. The negative is affixed to give a positive value

and the constant 10 assures that we get the Bohr radius as .529 in Angstroms when a value of N of the order of 40.5 is selected.

The solution in terms of the solar case must diverge and so inverts the atomic solution and requires a different boundary value.

$$R = \frac{-10\,c\sqrt{\dfrac{2}{3}}\,\sqrt{N+6}}{\alpha\,\ln(\alpha)}$$

The reason why the velocity of entities at greater distances in the solar system are less as opposed to the increased velocities of the atom

Radial Distribution

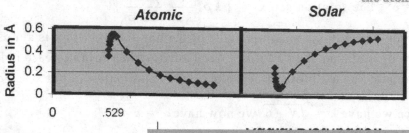

Kepler Did Not Know of Gravity

He observed that planets further away moved more slowly

Electrons further from the nucleus must move faster. This is true as the hydrogen radius increases when energy is added.

In atoms of more than one orbital, radii of each added orbital must be summed to find the radius of the atom.

"g" is for the superheavies.

$$\sum_{i}^{n}\sum_{l}^{j} r_i^{\,j}$$

n	s	p	d	f	g
X					
X	X				
X	X	X			
X					

Each mark (X) is a solution to the Bessel Equation

Sodium has a half less "orbital" than Calcium. Each / denotes one electron radius at that Attribute level, or N value.

	s	p	d	f	g
K	X				
L	X	X			
M	X	X			
N	X	\			
O					
P					

The electron has a higher frequency (greater velocity) in excited orbit. The only way to picture the electron further from the nucleus is to add its radial distance to its previous position.

A Planet has a lower frequency (i.e. mars 686 to 365 earth days to orbit the sun). Planetary orbits have divergent radii that require that x have the 1/kρ solution

Solutions for calcium {half orbital radii}

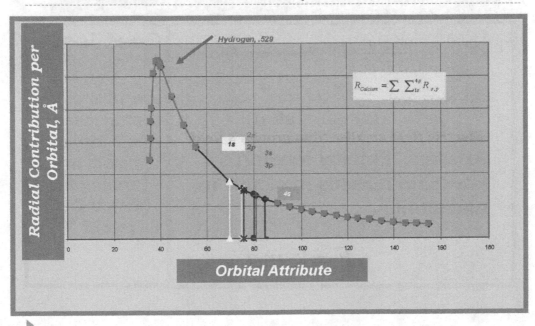

Principle Attributes Mirror the Radii

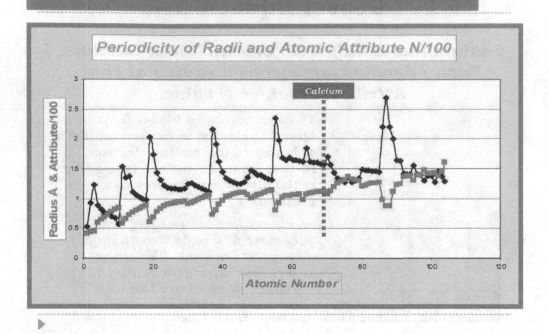

Periodicity of Radii and Atomic Attribute N/100

Electric and Gravitational Fields are Nearly Equal in the Atom.

Dirac: $\dfrac{KQq}{GMm} = 10^{39}$ **These are Forces**

In Reality $\dfrac{9x10^{10}(1.6x10^{-19})}{(6.6x10^{-11})1836\,(1)} = .011755$

Electric field smaller than gravity's field

Answer: Employ Newton's Law with masses in electron mass units.

$$F = k\,ma$$

Question:

Can one use 1836 as the central mass of the hydrogen atom in forming the Unified Field ???

In Your First Contact with Newton's Law You Perhaps Saw - -

$$F = k\, m\, a$$

And you were informed that the value of k is one when the unit kilogram is used

The value of K is also unity when the unit electron mass is used.

Again the Rydberg Constant can be Put in the Form Shown at Left Below:

$$\frac{n\hbar V}{q} = \frac{KQ\sqrt{\dfrac{mvr}{\hbar}}}{4\pi\varepsilon_o\,\varepsilon_r}$$

Now the second half is set equal to G since dividing by mass squared is legal ! !

I'll drink to that !!!

$$G = \frac{n\hbar V}{mm\,q}$$

| WLOG |

There was the lack of a radial solution to the wave equation, and no method to estimate charge density, ΔE, and this occasioned the "probability" solution of quantum mechanics

$$G = \frac{Q\sqrt{\dfrac{mvr}{\hbar}}}{4\pi\varepsilon_o\,\varepsilon_r M\,m} \qquad \nabla E = \frac{1}{\varepsilon_r}$$

$$K = \frac{1}{4\pi\varepsilon_o} \quad \text{and} \quad \sqrt{\frac{mvr}{\hbar}} = 1$$

$$\text{so } GM = \pm\frac{KQ\,\nabla E}{m}$$

Lacking only the radius to define the unified field of EM and Gravity

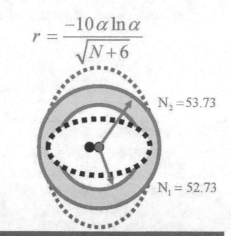

$$r = \frac{-10\,\alpha\ln\alpha}{\sqrt{N+6}}$$

$N_2 = 53.73$

$N_1 = 52.73$

One can assume that by varying orbit by up to a half an attribute, N, an ellipse will be formed

The center of the ellipse and the true position of the nucleus may not coincide.

Knowing the true position of the nucleus is not a priority

One must consider the time average radius to be that of the atom

$G = n\hbar V / Q$

Allows us to calculate Planetary Quantum Numbers

Planet	Perihelion	Mean	Aphelion
Mercury	1.92	2.12	2.36
Venus	2.88	2.89	2.90
Earth	3.37	3.40	3.43
Mars	4.0	4.18	4.39
Jupiter	7.72	7.75	7.75
Saturn	10.16	10.44	10.74
Uranus	15.46	14.89	15.25
Neptune	18.51	18.62	18.73
Pluto	19.16	21.55	24.63

The concept of trying to reconcile Quantum Mechanics with General Relativity is oftentimes referred to as the *problem of quantum gravity*. <u>Leon Lederman</u> has said that the system we have now is just too complex. <u>John Archibald Wheeler</u> remarked that once we see the solution that "it will be so beautiful that we will all say how could it have been otherwise"? It is almost axiomatic that something simply has been overlooked

▶

Kepler's Law requires that the mass ratio be central mass to unit mass.

$$K = \frac{G\,\{\text{central mass}\}}{\{\text{unit mass}\}}$$

Mass in Kilograms, Ounces, or in electron masses

For the atom $K = 1836\,G$

Johannes Kepler

▶

It is axiomatic, if this is true, that there should be a dimensional transformation of the atomic field to reflect this. Unfortunately when this is done and **_G is changed to electron mass (EM) across the equal sign_** the dimensional analysis is somewhat unconventional.

$$\{Physical\ unity\ is\ represented\ by\ \frac{(1.1\times 10^{30})^2 EM^2}{kg^2}$$

Unconvincing

$$F\ \frac{kg\ meters}{sec^2} = G\ \frac{kg\ meters^3}{sec^2}\ \frac{1.7x10^{-27}\ kg\ x\ 9.1x10^{-31}kg}{R^2\ meters^2}\ \frac{(1.1\times 10^{30})^2 EM^2}{kg^2}$$

$$F\ \frac{EM\ meters}{EM\ sec^2} = G\ \frac{EM\ meters^3}{EM^2 sec^2}\ \frac{1836\ EM}{R^2\ meters^2}$$

$$WLOG \qquad g = F\ \frac{meters}{sec^2} = \frac{1836\ G\ meters}{sec^2}$$

Galileo Demonstrated that Unit Gravity was Relative to Central Mass.

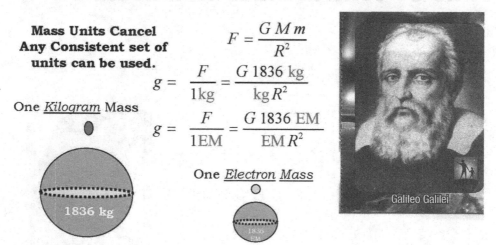

Mass Units Cancel Any Consistent set of units can be used.

One <u>Kilogram</u> Mass

1836 kg

$$F = \frac{G\ M\ m}{R^2}$$

$$g = \frac{F}{1kg} = \frac{G\ 1836\ kg}{kg\ R^2}$$

$$g = \frac{F}{1EM} = \frac{G\ 1836\ EM}{EM\ R^2}$$

One <u>Electron Mass</u>

1836 EM

Galileo Galilei

For Earth 6 x 1024 <u>Kilograms</u> *or* in the Atom 1836 <u>Electron Masses</u>

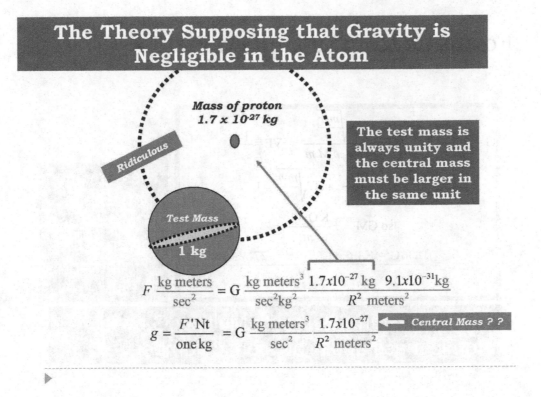

The Theory Supposing that Gravity is Negligible in the Atom

Mass of proton
1.7 x 10⁻²⁷ kg

Ridiculous

The test mass is always unity and the central mass must be larger in the same unit

Test Mass

1 kg

$$F \frac{kg \ meters}{sec^2} = G \frac{kg \ meters^3}{sec^2 kg^2} \frac{1.7x10^{-27} \ kg \quad 9.1x10^{-31} kg}{R^2 \ meters^2}$$

$$g = \frac{F'Nt}{one \ kg} = G \frac{kg \ meters^3}{sec^2} \frac{1.7x10^{-27}}{R^2 \ meters^2} \longleftarrow Central \ Mass \ ? \ ?$$

It is axiomatic, if this is true, that a transformation of the atomic field is also possible to reflect this. Accordingly we shall employ the physical unitary transformation ratio of the number of electron masses per kg.

{*Physical unity is represented by* $\frac{(1.1x \ 10^{30})^2 EM^2}{kg^2}$

$$F \ EM^2 \frac{meters}{sec^2} = G \frac{kg \ meters^3}{sec^2} \frac{1.7x10^{-27} \ kg \ x \ 9.1x10^{-31} kg}{R^2 \ meters^2} \frac{(1.1x \ 10^{30})^2 EM^2}{kg^2}$$

$$F \ EM^2 \frac{meters}{9.1x10^{-31} \ sec^2} = G \frac{EM \ meters^3}{sec^2} \frac{1.7x10^{-27}}{R^2 \ meters^2} \frac{(1.1x \ 10^{30})^2 EM}{}$$

$$F \frac{EM \ meters}{9.1x10^{-31} \ 1.1x \ 10^{30} \ EM \ sec^2} = G \frac{EM \ meters^3}{EM \ sec^2} \frac{1.7x10^{-27} \ x1.1x \ 10^{30}}{R^2 \ meters^2}$$

$$WLOG \qquad g = F \frac{meters}{sec^2} = \frac{1836 \ G \ Meters}{sec^2}$$

Formation of the Unified Field

$$G = \frac{n\hbar V}{mm\,q} \qquad G = \frac{Q\sqrt{\frac{mvr}{\hbar}}}{4\pi\varepsilon_o\,\varepsilon_r M\,m} \qquad \nabla E = \frac{1}{\varepsilon_r}$$

$$K = \frac{1}{4\pi\varepsilon_o} \quad \text{and} \quad \sqrt{\frac{mvr}{\hbar}} = 1$$

$$\text{so } GM = \pm\frac{K\,Q\,\nabla E}{m}$$

$$g_{unified} = \frac{1836\,G + K\,1.6\times10^{-19}(9N^2 - 271N + 586)/49}{R^2}$$

The derivation of ΔE, the Poisson, is given in the following slide.

For succeeding elements Multiply G by atomic mass and K by atomic number.

$$\nabla E = (N+5)\,F + 2\,F'$$

For any value of N, the Poisson can be evaluated as a scalar polynomial by substituting the value of F in the equation below and solving for α.

$$\nabla^2\phi = (N+5)F + 2F' - (N-28)F^{-1} + 7 \qquad F = \frac{1}{1-\alpha}$$

$$\nabla^2\phi = \frac{(9N^2 - 271N + 588)}{49} \qquad \alpha = \frac{N-35}{N-28} \qquad R = -\frac{10\,\alpha\ln(\alpha)}{\sqrt{N+6}}$$

Forming Fields from the Rydberg Equation

$$G = \frac{n\hbar V}{2\pi q} \neq \frac{KQ\sqrt{mvr}}{\{M\ m\}}$$

$$\frac{G M}{R^2} = \frac{KQ(\pm\sqrt{1})}{mR^2\ \varepsilon_{rel}}$$

**Units
Certainly
Agree**

Rydberg Version of the Unified Field

**We can now add *relative permittivity*
which when identified with the
Poisson accomplishes the equality.**

Why aren't the Attributes Integers ? ?

$$\sqrt{\frac{mvr}{h}}$$ Has the distinction of always being ± 1

Jumps = n + ℓ

n Principle Quantum Number	Orbital Quantum Number	ℓ s 0	p	d	f	g
1		1				
2			6	7		
3			11	12	13	
4			16			
5						
6						

Result: We have Integral Quantum Numbers; but They are Imposed Upon Real Numbered Attributes

	Attributes N		
1	49.6		
2	50.6		
3	54.6	55.6	56.6
4	59.6	60.6	
5	s	p	d

In the solar system the value of GM equals KQq/m. It can be shown that Q is the central charge i.e. the central mass x the positive charge per kg and q/m is the negative charge per kg.

We now have a _Multiplicative way_ to equalize all H values by means of a relative permittivity due to the gases the photon must traverse.

Positive charge to mass ratio of a kilogram

$$Q^+/m = 1.602 \times 10^{-19}(6.02 x 10^{26})$$

$$= 9.58 x 10^7$$

Negtive charge to mass ratio of any planet

$$q^- / m_{planet} = 7075 \, x 10^{-29} \quad \text{coul/kg}$$

$$\frac{GM}{R} = \frac{K\,M_{Sun}\,kg(Q^+ coul\,/\,kg \,\,x\,\, q^- coul/kg)}{R}$$

The possibility exists that G may not have the same value everywhere

$$\frac{G\,M_{Sun}}{R} = \frac{M_{Sun}\,8.98 \times 10^9 \,\, 9.58 \times 10^7 \,\, 7.75 \,x10^{-29}\,coul}{R \, kg \,\,\, \varepsilon_{Relative}}$$

$$G = 8.98 \times 10^9 \,\, 9.58 \times 10^7 \,\, 7.75 \,x 10^{-29}$$

$$= 6.672 \,x 10^{-11}/\varepsilon_{Relative}$$

Completing the "Package"

Assume we now have the solution to Schrödinger's Equation, the Unified Field, we can Show One Force in Nature and that Dark Matter has been Explained.

There are still three other "icons" of the past to tackle

Dark Energy

Redshift

"Constant" Hubble or Density

Dark Energy

Dark Energy may be an electromagnetic imbalance at the outer edge of the universe

Electrons under the impact of the original Big Bang may have populated the outer edges in greater numbers creating an attraction for the slower moving mass.

There may be no Dark Energy at all !!

No Dark Energy ? ?

Perhaps the outward movement is just an artifact of the Redshift Phenomenon

The Doppler concept applied to a photon particle is not exactly as it is applied to a sound wave. Both have variable wavelengths but:

Light waves have constant speed and vary wavelength and frequency over a many light years.

Passing sound waves vary all three components at a single instant.

Sound Experiences a Deceleration with Concomitant Frequency Loss.

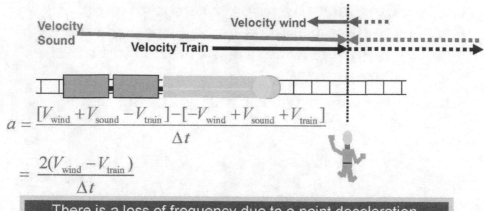

$$a = \frac{[V_{wind} + V_{sound} - V_{train}] - [-V_{wind} + V_{sound} + V_{train}]}{\Delta t}$$

$$= \frac{2(V_{wind} - V_{train})}{\Delta t}$$

There is a loss of frequency due to a point deceleration.

Contrast the slow loss of the frequency of light as the quantum overcomes an increasing field of gravity.

It is the frequency of the photon ⬭ (not the frequency of photons from a receding object) that is relevant.

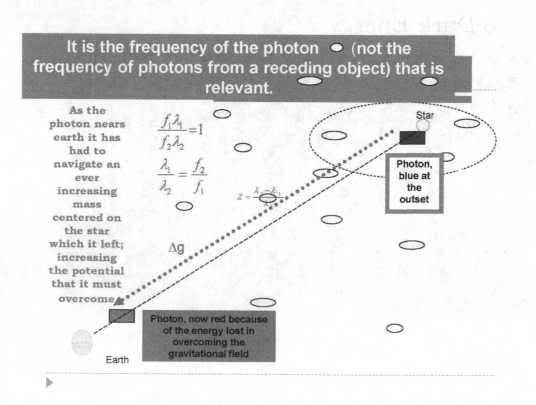

As the photon nears earth it has had to navigate an ever increasing mass centered on the star which it left; increasing the potential that it must overcome

$$\frac{f_1 \lambda_1}{f_2 \lambda_2} = 1$$

$$\frac{\lambda_1}{\lambda_2} = \frac{f_2}{f_1}$$

$$z = \frac{\lambda_2 - \lambda_1}{\lambda}$$

Δg

Star

Photon, blue at the outset

Photon, now red because of the energy lost in overcoming the gravitational field

Earth

Hubble's Constant and Quantum Numbers of the Nearest Stars

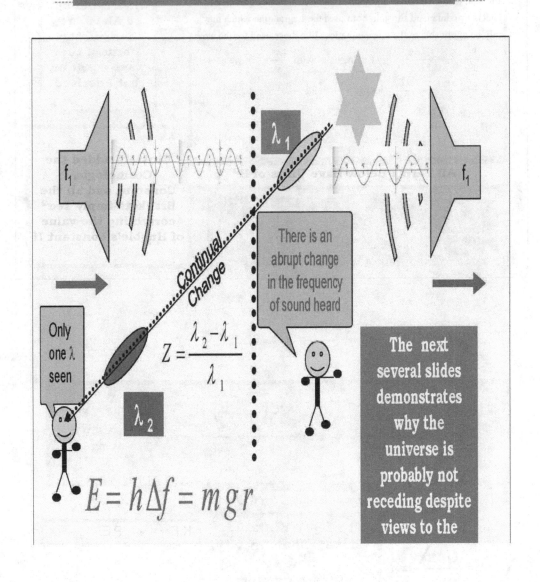

Friedman's solution to Einstein's General Relativity

Derivation of the Friedman equation, equivalent to Einstein's spherical expression of General Relativity, is done by assuming that the sum of the kinetic and potential energies is locally conserved. That is Friedman and Einstein both used the Lagrangian which may only be applicable to the wave equations like those used for the atom.

$$\frac{1}{2}mV^2 - \frac{GMm}{R} = E$$ Replacing M with $\frac{4\pi}{3}R^3\rho$ we obtain

$$V^2 = \frac{8\pi G}{3}\rho + \frac{2E}{mR^2}$$ letting $\frac{2E}{m}$ equal $-kc^2$ we recover the

Friedman equation:

$$\left(\frac{dR/dt}{R}\right)^2 = \frac{8\pi G\rho}{3} - \frac{kc^2}{R^2}$$

All Terms Below Have Units of H²

$$\left(\frac{dR/dt}{R}\right)^2 = \frac{8\pi G\rho}{3} - \frac{kc^2}{R^2} \pm \frac{\Lambda}{3}$$

The Lagrangian was involved in the 1770's when Euler and d'Alembert derived the form of the wave equation that I derived.

Einstein added the Cosmological Constant and all the Brackets imply sec⁻² correcting the value of Hubble's constant H

$$\left(\frac{dR/dt}{R}\right)^2 = \frac{8\pi G\rho}{3} - \frac{kc^2}{R^2} \pm \frac{\Lambda}{3}$$

All Terms above Have Units of H²

$\frac{m}{} = E$

K.E   -   P.E

In 1937 Einstein Stated That Gravity had an Electromagnetic Solution in a *Spherical* Universe ! ! !

$$mV^2 = \frac{GMm}{R} = E$$

The velocity on the right is the escape velocity

The starting Lagrangian

My Difficulty with the Lagrangian II

$$\frac{1}{2}mV^2 - \frac{GMm}{R} = E \quad \Longleftrightarrow \quad mV^2 = \frac{GMm}{R} = E$$

$$V^2 = \frac{GM}{R}$$

$$V^2 = \frac{GM\dfrac{4\pi}{3}R^2}{\dfrac{4\pi}{3}R^3}$$

Defining the Hubble Constant from Newton's Law

$$\frac{V^2}{R^2} = \frac{4\pi G\rho}{3}$$

$$H^2 = \frac{4\pi G\rho}{3}$$

There seems here a confusion between velocity of free fall and velocity of entities in orbit. The Hamiltonian, if used in the latter regard, would certainly be true for the total energy **in free fall**. One cannot say anything for the Lagrangian here except that the energy of escape exceeds the energy in orbit. I do not think escape energy is an appropriate mechanism, **_for cosmology_**. Although, in the atom it could define the transition state width at an appropriate V.

The Clincher

The Velocity used in the definition of the redshift is radial; and, it is not just directed outward.

V cannot be radial and only directed outward.

$$\frac{V^2}{R^2} = \frac{4\pi G\rho}{3}$$

V can only be tangential and radial because it is part of Newton's Law ! !

$$\frac{1}{2}MV^2 - MgR = 0$$

$$V^2 = 2gR$$

$$V = \sqrt{2gR}$$

$$\frac{dR}{dt} = \sqrt{2gR}$$

The velocity is the escape velocity. This phenomenon was presented by the Bernoulli's, Leibnitz, Newton, and l'Hospital

It certainly applies in orbit to characterize an ellipse which would occur in any isolated "jump" system about a central Mass.

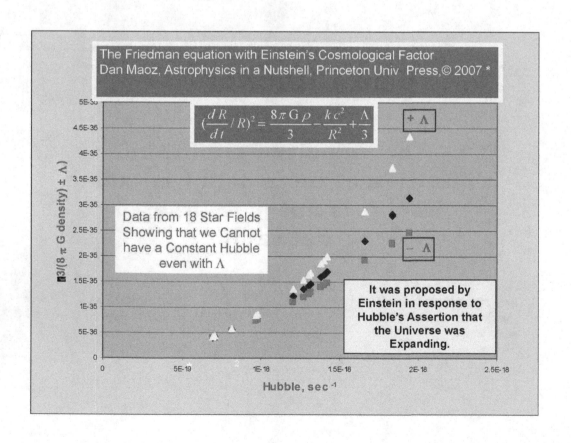

The Friedman equation with Einstein's Cosmological Factor
Dan Maoz, Astrophysics in a Nutshell, Princeton Univ Press,© 2007 *

$$\left(\frac{dR}{dt}/R\right)^2 = \frac{8\pi G\rho}{3} - \frac{kc^2}{R^2} + \frac{\Lambda}{3}$$

$+ \Lambda$

$- \Lambda$

Data from 18 Star Fields Showing that we Cannot have a Constant Hubble even with Λ

It was proposed by Einstein in response to Hubble's Assertion that the Universe was Expanding.

■3/(8 π G density) ± Λ

Hubble, sec $^{-1}$

Can We Define an Ultimate Density of the Universe ? ?

$$V^2 = \frac{GM}{R}$$

$$V^2 = \frac{GM\frac{4\pi}{3}R^2}{\frac{4\pi}{3}R^3}$$

$$\frac{V^2}{R^2} = \frac{4\pi G\rho}{3}$$

$$H^2 = \frac{4\pi G\rho}{3} \Longleftrightarrow \left(\frac{dr/dt}{dr}\right)^2 = \frac{4\pi G\rho}{3}$$

$$E = mc^2$$

$$\frac{M}{V} = \frac{E\infty}{V\infty c^2}$$

$$\rho = \frac{1}{c^2}$$

$$Conclusion:\ E\infty < V\infty$$

Friedman's Equation
$$\left(\frac{dR/dt}{R}\right)^2 = \frac{8\pi G\rho}{3} - \frac{kc^2}{R^2}$$

The density from Einstein's equation gives too great a value and implies much too low value of the Hubble value. Radii imply volumes and the variable k in Friedman's equation is arbitrary to the point where the velocity Z^2c^2 of the Redshift Z can be used to specify Hubble's constant from Newton's Law. This implies that the 8 $\pi G\rho/3$ term of Einstein's equation is double the Hubble.

The Density given by $1/c^2$ is too High

Next assume $m_o c^2$ is the mass tied up in pure energy then the mass remaining as matter energy is mcv

$$E = mc^2$$

$$\frac{M}{V} = \frac{E\infty}{V\infty c^2}$$

The infinities do not cancel equally

$$\rho = \frac{1}{c^2} Conclusion: E\infty < V\infty$$

Consider the Einstein's Energy Squared Formula
$$E^2 = m_o^2 c^4 + m^2 c^2 v^2$$

How does one assess the infinity of mass against the infinity of area

if $E_m = m c v$

$$m = \frac{E_m}{c^2}\ and\ \ \rho = \frac{mcv}{c^2 V}$$

The units on the right are

$$\frac{mcR/T}{cc(4\pi/3)RRR}\ or\ \frac{m}{cRR}\ ie\ \frac{kg}{c\ meters^2}$$

Clearly there are different orders of infinity between kilograms and meters

Consider the Following

$$if\ E_m = m\ c\ v$$

$$m = \frac{E_m}{c^2}\ and\ \ \rho = \frac{mcv}{c^2 V} = \frac{m\ v}{cV}$$

Assume canceling infinities of mv and volume

Also assume 6.02×10^{26} *H atoms per Kg and*

$$1/2(.529 \times 10^{-10})\ H\ atoms\ per\ meter^3$$

$$\rho = \frac{\propto Kg}{6.02 \times 10^{26}\ atoms}\ \frac{264332\sqrt{2}\ (m/\sec)\ 9.45 \times 10^9\ atoms}{\propto (1meter)^3\ 299792458}$$

$$=1.96 \times 10^{-20}\ kg\ /\ meter^3$$

$$H^2 = \frac{4\pi G \rho}{3}$$

$$= \frac{4\pi\ 6.672 \times 10^{-11}\ 1.96 \times 10^{-20}}{3}$$

$$H = 2.34 \times 10^{-15}\ or\ 71.6 \pm 3\ km/\sec/mpc$$

An Article from the Encyclopedia Britannica Represents an Opinion Regarding the Fate of the Universe RE: (1..25 E-25) my Predicted of the Previous Slide

      If Hubble's constant at the present epoch is denoted as *H*0, then the closure density (corresponding to an Einstein–de Sitter model) equals 3*H*02/8π*G*, where *G* is the universal gravitational constant in both Newton's and Einstein's theories of gravity. The numerical value of Hubble's constant *H*0 is 22 kilometers per second per million light-years; the closure density then equals 10^{-26} kilogram per cubic meter, the equivalent of about six hydrogen atoms on average per cubic meter of cosmic space. If the actual cosmic average is greater than this value, the universe is bound (closed) and, though currently expanding, will end in a crush of unimaginable proportion. If it is less, the universe is unbound (open) and will expand forever. The result is intuitively plausible since the smaller the mass density, the smaller the role for gravitation, so the more the universe will approach free expansion (assuming that the cosmological constant is zero).

$$R_{excess} = \sqrt{\frac{A}{4\pi}} - R_{measured} = \frac{GM}{3c^2}$$

$$0 = \sqrt{\frac{A}{4\pi}} - R_{measured} = \frac{G\rho 4\pi R^3_{measured}}{9c^2}$$

$$= \sqrt{\frac{4\pi R}{4\pi R_{meas}}} \; x \; \frac{1}{R^2_{meas}} - \frac{1}{R^3_{measured}} = \frac{G\rho 4\pi}{9c^2}$$

$$= \frac{1}{R^2_{meas}} = \frac{G\rho 4\pi}{3c^2} = 7.77426 \times 10^{-53}$$

$R = 1.1341499 \times 10^{26}$ Meters   *Volume* $= 6.1108 \times 10^{78}$ M^3

Using the Density of the Slide Before

The Mass Becomes

7.63×10^{52} Kg

Now knowing the mass and radius we can calculate the radius excess to find Einstein's equation for curvature.

Using the Predicted Mass of the Universe

$$\text{Radius excess} = \sqrt{\frac{A}{4\pi}} - r_{measured} = \frac{GM}{3c^2}$$

Formula from *Six Not-So-Easy Pieces*, Richard P. Feynman Perseus Books, Reading, Mass. © 1997

$$= \frac{6.672\,x10^{-11}\,x\,7.63\,x10^{52}}{3(299792458)^2}$$

$$R_{excess} = 1.889\,x10^{25}$$

$$R = 1.1341499 \times 10^{26} \text{ Meters}$$

Total R + excess 1.32205 x 10^{26} *meters*

Curvature 7.56401 x 10^{-27} (*near zero*)

CONTRAST EARTH'S RADIAL EXCESS 1 mm

Left - an animation showing the brightest galaxies (AMag<-16.3) in the Virgo supercluster. This animation is an actual plot of the 2500 most luminous galaxies within 100 million light years from us. The slight uncertainty of the distances to many of these galaxies makes the galaxies appear more scattered than they actually are, but the general features of the Virgo supercluster are obvious, especially the long filaments of galaxies and the low-density void regions.

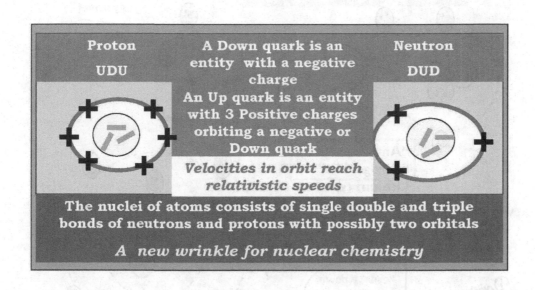

Proton
UDU

A Down quark is an entity with a negative charge

An Up quark is an entity with 3 Positive charges orbiting a negative or Down quark

Velocities in orbit reach relativistic speeds

Neutron
DUD

The nuclei of atoms consists of single double and triple bonds of neutrons and protons with possibly two orbitals

A new wrinkle for nuclear chemistry

Neutron   **And Also**   Proton

Using q/3 charge entities a proton has 6 positive orbiters of 3 negative downquark centers **whereas a neutron has only three.**

This allows for multiple bond bonding and relativistic mev increases that mimic the masses of the particle zoo.

If you think of frequencies of curled up time averages as *time dimensions* **there are a multiplicity (>16) of** *strings.*

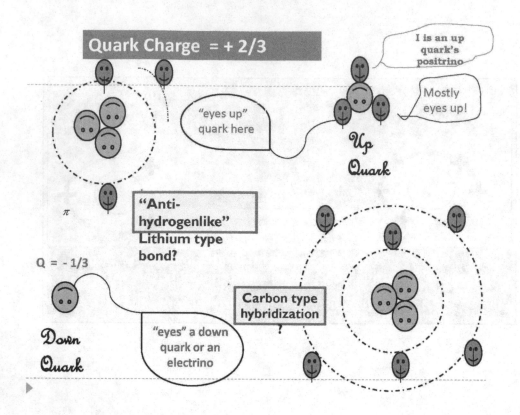

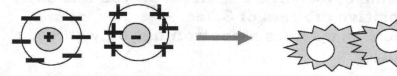

Assume that the pair production produces two differently charged bodies, matter and antimatter with 12 (1/3) q charges in orbit

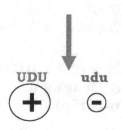

What is left after the contact is a proton and an electron impelled into orbit by the annihilation of the outer rings.

One has to *further assume* that the positive interior always receives the mass lost by the negative; else, we would have a mixture of matter and antimatter all over again.

udu

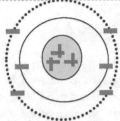

The electron behaves a tiny anti-particle. It's negative q charge breaks into six q/3's orbiting an inner three positive q/3's

It can be shown that in the deBroglie relation, $n\hbar = mvr$ the mass is relativistic to one riding the charge passing the mass associated with the charge at the speed of light. The mass orbits more nearly at the speed of sound while its charge tracks at c.

Question:

How is the neutron formed ?

Wrap up: Also - AM - Can do

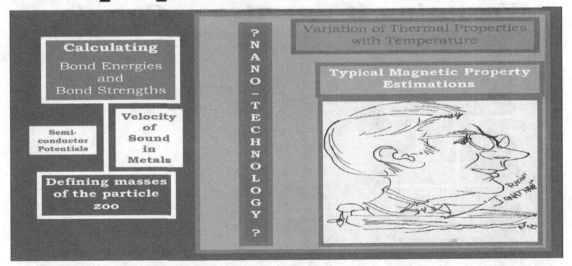

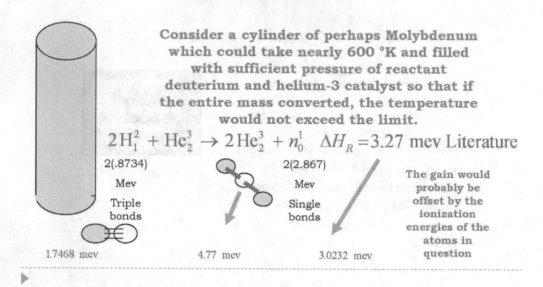

A Plea for a Cooler Fusion

Consider a cylinder of perhaps Molybdenum which could take nearly 600 °K and filled with sufficient pressure of reactant deuterium and helium-3 catalyst so that if the entire mass converted, the temperature would not exceed the limit.

$$2\,H_1^2 + He_2^3 \rightarrow 2\,He_2^3 + n_0^1 \quad \Delta H_R = 3.27 \text{ mev Literature}$$

2(.8734)
Mev
Triple bonds

2(2.867)
Mev
Single bonds

The gain would probably be offset by the ionization energies of the atoms in question

1.7468 mev

4.77 mev

3.0232 mev

Wrapping it up: Also - - Can do

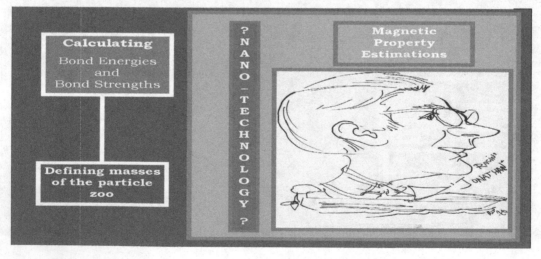

Index